Advances in Plant Biotechnology

The goal of *Advances in Plant Biotechnology* is to integrate the most recent knowledge on plant tissue culture, secondary metabolites production under controlled conditions, scaling up to produce them at bioreactor level, and their industrial applications. The biosynthetic pathways and the factors that affect them and the accumulation of metabolites, including metabolomics in medicinal plants, are key components as well. Several extraction and encapsulation technological procedures are reviewed. The structure and function of metabolites from selected commercial crops are reported in detail. Finally, items of paramount importance, such as bioavailability and stability of metabolites in pharma and food products are deeply analyzed.

Key Features:

- Strategies for obtaining selected metabolites through *in vitro* culture.
- Application of biotechnological and bioengineering principles to the management of plant metabolites.
- Description of the encapsulation of selected metabolites.
- Bioavailability and stability of metabolites in pharma, food, and industrial sectors.

This book is mainly addressed to research scientists, technical staff, and private and public organizations involved in plant biotechnology and its processing industries. Last, but not least, students at all levels and post-doctoral researchers have received special attention from all editors and authors in this publication.

Food Biotechnology and Engineering

Series Editor: Octavio Paredes-López

Volatile Compounds Formation in Specialty Beverages
Edited by Felipe Richter Reis and Caroline Mongruel Eleutério dos Santos

Native Crops in Latin America
Biochemical, Processing, and Nutraceutical Aspects
Edited by Ritva Repo-Carrasco-Valencia and Mabel C. Tomás

Starch and Starchy Food Products
Improving Human Health
Edited by Luis Arturo Bello-Perez, Jose Alvarez-Ramirez, and Sushil Dhital

Bioenhancement and Fortification of Foods for a Healthy Diet
Edited by Octavio Paredes-López, Oleksandr Shevchenko, Viktor Stabnikov, and Volodymyr Ivanov

Latin-American Seeds
Agronomic, Processing and Health Aspects
Edited by Claudia Monika Haros, María Reguera, Norma Sammán, and Octavio Paredes-López

Advances in Plant Biotechnology
In Vitro Production of Secondary Metabolites of Industrial Interest
Edited by Alma Angélica Del Villar-Martínez, Juan Arturo Ragazzo-Sánchez, Pablo Emilio Vanegas-Espinoza, and Octavio Paredes-López

Bioconversion of Wastes to Value-added Products
Edited by Olena Stabnikova, Oleksandr Shevchenko, Viktor Stabnikov and Octavio Paredes-López

For more information about this series, please visit: www.routledge.com/Food-Biotechnology-and-Engineering/book-series/CRCFOOBIOENG

Advances in Plant Biotechnology

In Vitro Production of Secondary Metabolites of Industrial Interest

Edited by
Alma Angélica Del Villar-Martínez,
Juan Arturo Ragazzo-Sánchez,
Pablo Emilio Vanegas-Espinoza, and
Octavio Paredes-López

CRC Press
Taylor & Francis Group
Boca Raton London New York

CRC Press is an imprint of the
Taylor & Francis Group, an **informa** business

Designed cover image: © Shutterstock

First edition published 2024
by CRC Press
6000 Broken Sound Parkway NW, Suite 300, Boca Raton, FL 33487-2742

and by CRC Press
4 Park Square, Milton Park, Abingdon, Oxon, OX14 4RN

CRC Press is an imprint of Taylor & Francis Group, LLC

Library of Congress Cataloging-in-Publication Data

Names: Villar-Martinez, Alma Angelica Del, editor. | Ragazzo-Sanchez, Juan Arturo, editor. | Vanegas-Espinoza, Pablo Emilio, editor. | Paredes-Lopez, Octavio, editor.
Title: Advances in plant biotechnology : in vitro production of secondary metabolites of industrial interest / edited by Alma Angelica Del Villar-Martinez, Juan Arturo Ragazzo-Sanchez,
Pablo Emilio Vanegas-Espinoza, and Octavio Paredes-Lopez.
Other titles: In vitro production of secondary metabolites of industrial interest | Food biotechnology series.
Description: First edition | Boca Raton, FL ; London : CRC Press, 2024 |
Series: Food biotechnology and engineering | Includes bibliographical references and index. | Also available online.
Identifiers: LCCN 2023004152 (print) | LCCN 2023004153 (ebook) | ISBN 9781000923520 (epub) | ISBN 9781003166535 (ebook) | ISBN 9780367746926 (hardback) | ISBN 9780367763381 (paperback)
Subjects: LCSH: Plant biotechnology. | Plant metabolites. | Metabolism, Secondary.
Classification: LCC QK887 (ebook) | LCC QK887 .A38 2024 (print) |
DDC 572/.42 23/eng/20230--dc30
LC record available at https://lccn.loc.gov/2023004152

ISBN: 9780367746926 (hbk)
ISBN: 9780367763381 (pbk)
ISBN: 9781003166535 (ebk)

DOI: 10.1201/9781003166535

Typeset in Minion
by Deanta Global Publishing Services, Chennai, India

Contents

Editors

Alma Angélica Del Villar-Martínez, PhD, is a Scientist at the Molecular Biology Laboratory in the Department of Biotechnology at Centro de Desarrollo de Productos Bióticos (CEPROBI) in the Instituto Politécnico Nacional (IPN) in Morelos, Mexico. She received her bachelor's degree in Bio-pharmaceutical Chemistry, MSc, and PhD in Plant Biotechnology from the Department of Biotechnology at CINVESTAV-Irapuato, Mexico, as well as being a member of the National System of Researchers (Sistema Nacional de Investigadores, SNI-CONACYT). She supervises the research of undergraduate/graduate students and visiting scholars and works closely with local and international research. Dr. Del Villar-Martínez carried out research projects related to the study of medicinal plants with pharmacological potential. She has been involved in evaluating the biological potential of extracts from plant sources, in the development and induction of compounds with an antioxidant and anticancer effect, in the analysis of the chemical profile of extracts from different medicinal plants, the study of pigments of pharmacological and food interest, and the use of different biotechnological tools to improve the biosynthesis of the metabolites of interest from *in vitro* cultures. She aims to generate knowledge concerning the development of products of industrial interest for their application in biological systems and their use in health care. Dr. Del Villar-Martínez has published 25 peer-reviewed research papers, two book chapters, among other reviews and abstracts.

Juan Arturo Ragazzo-Sánchez, PhD, is a Professor-Researcher at Tecnológico Nacional de México/Instituto Tecnológico de Tepic, in the Biotechnology and Food Engineering, Department of Food Science. He supervises research of undergraduate/graduate students and visiting scholars and work closely with local and international research, education, and international programs (CYTED, European Union) and partnerships with the private sector. He is a graduate in Chemical Engineering and has an MSc. in Biochemical Engineering, both from Instituto Tecnológico de Veracruz. Later, he received his PhD in Food Sciences from the Doctoral School of Biological and Industrial Sciences and Processes (University of Montpellier, France). He has also been a visiting researcher at different institutions, including Novel Materials and Nanotechnology Group of IATA-CSIC, Spain under the tutelage of Prof. Jose Maria Lagaron and at the Department of Génie Biologique et Agroalimentaire – UMR IATE – Université de Montpellier, France; under the tutelage of Mme. Prof. Dominique Chevalier-Lucia.

The major theme in Dr. Ragazzo-Sanchez's research is the utilization of different strategies to improve the extraction of high-value biological compounds (HVBC) from plant sources and their protection by encapsulation. These strategies include the use of different extraction methods, and micro- and nano-encapsulation processes, as well as the development of new biopolymer materials obtained from natural sources to be used in encapsulation processes for HVBC, enzymes, and antagonistic microorganisms to achieve formulations used in the control of phytopathogens in pre- and post-harvest stages in agriculture. Dr. Ragazzo-Sánchez has published 75 peer-reviewed research papers and 12 book chapters, edited one book, and published many other reviews and abstracts. He has also led more than 50 research projects and been awarded five patents. In 2019, he was Co-Coordinator of the DESEAAR-CYTED Network, 319RT0576 (Desarrollo Sostenible en Agroalimentación y Aprovechamiento de Residuos industriales).

Pablo Emilio Vanegas-Espinoza, PhD, is a Researcher at Centro de Desarrollo de Productos Bióticos (CEPROBI) of Instituto Politécnico Nacional (IPN), in the Molecular Biology Laboratory of the Department of Biotechnology. As a professor, he teaches subjects such as Molecular Biology, Selected Topics on Secondary Metabolites in Plants, Biotechnology of *In Vitro* Cultures; he also supervises research of undergraduate/graduate students. He is an Agricultural Engineer in Horticulture and was awarded his MSc and PhD in Plant Biotechnology from Centro de Investigación y de Estudios Avanzados (CINVESTAV-IPN), Campus Guanajuato, Mexico. Dr. Vanegas-Espinoza has worked on plant cell culture and genetic transformation for more than 25 years.

Octavio Paredes-López obtained his Bachelor's degree in Biochemical Engineering, and a Master in Food Science and Technology at the National Polytechnic Institute in Mexico City; a Master in Biochemical

Engineering at the Czech Academy of Sciences; and his PhD at the University of Manitoba, Dept. of Plant Sciences, Winnipeg, Canada. He has completed postdoctoral and research visits in several universities and institutes in countries like the USA, Canada, France, the UK, Germany, Switzerland, and Brazil. He has published over 245 scientific papers in refereed international journals, 55 reviews and book chapters, is author/editor/coeditor of ten books, and author of 125 articles on the role of science in popular newspapers mainly in Mexico and some in the USA and France; in total over 435 works. His scientific publications have been on basic and applied aspects of foods (i.e, fruits, cereals, plant foods, molecular biology, GMO's); on microbial topics (i.e., genetic modifications, fermentation technology, overexpression of secondary compounds); and on biotechnology. Over 100 students have graduated, doing research in his laboratory, some of them from abroad. He has received all Mexican awards in his area of research including the highest recognition: Mexican Award in Science. Emeritus Professor, and Emeritus National Researcher. International recognition: Institute of Food Technologists Fellow USA; WK Kellogg International Food Security Award; Nestlé Award; Academy of Sciences of the Developing World-Agriculture Award, Trieste, Italy. Ordre National du Mérite, Republique Française, Paris, France. Vice-President and President of the Mexican Academy of Sciences. Doctor Honoris causa of three Mexican universities. Doctor Honoris causa of the University of Manitoba, Canada, where he obtained his PhD.

Contributors

A. Berenice Aguilar-Guadarrama, Centro de Investigaciones Químicas, IICBA, Universidad Autónoma del Estado de Morelos, Mexico.

Gabriela Aguilar-Hernández, Tecnológico Nacional de México, Instituto Tecnológico de Tepic, Nayarit, Mexico.

Maria Guadalupe Aguilar-Uscanga, Tecnológico Nacional de México, Instituto Tecnológico de Veracruz, Mexico.

Luis Miguel Anaya-Esparza, Centro Universitario de los Altos, Universidad de Guadalajara, Jalisco, Mexico.

Yessica Casales-Tlatilpa, Instituto Politécnico Nacional, Centro de Desarrollo de Productos Bióticos, Morelos, Mexico.

Andrés Chavez-Salazar, Universidad de Caldas, Departamento de Ingeniería, Facultad de Ingenierías, Manizales, Caldas, Colombia.

Carla Norma Cruz-Salas, Tecnológico Nacional de México, Instituto Tecnológico de Tepic, Nayarit, Mexico.

Alma Angélica Del Villar-Martínez, Instituto Politécnico Nacional, Centro de Desarrollo de Productos Bióticos, Morelos, Mexico.

Javier Gómez-Rodríguez, Tecnológico Nacional de México, Instituto Tecnológico de Veracruz, Mexico.

Elda Margarita González-Cruz, Tecnológico Nacional de México, Instituto Tecnológico de Tepic, Nayarit, Mexico.

Katia Nayely González-Gutiérrez, Tecnológico Nacional de México, Instituto Tecnológico de Tepic, Nayarit, Mexico.

Ana Silvia Gutiérrez-Román, Instituto Politécnico Nacional, Centro de Desarrollo de Productos Bióticos, Morelos, Mexico.

Javier Darío Hoyos-Leyva, Fundación Universitaria Agraria de Colombia, Facultad de Ingeniería Agroindustrial, Grupo de Investigación e Innovación Agroindustrial - GINNA, Bogotá D.C., Colombia.

Lina María Londoño-Giraldo Universidad Libre Seccional Pereira, Risalda, Colombia.

Brandon Alexis López-Romero, Tecnológico Nacional de México, Instituto Tecnológico de Tepic, Nayarit, Mexico.

Edmundo Lozoya-Gloria, Centro de Investigación y de Estudios Avanzados-IPN, Unidad Irapuato, Guanajuato, Mexico.

Iván Luzardo-Ocampo, Instituto de Neurobiología, Universidad Nacional Autónoma de México (UNAM) - Campus Juriquilla, Querétaro, Mexico.

Efigenia Montalvo-González, Tecnológico Nacional de México, Instituto Tecnológico de Tepic, Nayarit, Mexico.

Norma Elizabeth Moreno-Anzúrez, Instituto Politécnico Nacional, Centro de Desarrollo de Productos Bióticos, Morelos, Mexico.

Yolanda Nolasco-González, Tecnológico Nacional de México, Instituto Tecnológico de Tepic, Nayarit, Mexico.

Octavio Paredes-López, Centro de Investigación y de Estudios Avanzados-IPN, Irapuato Guanajuato, Mexico.

Cristina Prieto López, Grupo de Nuevos Materiales y Nanotecnologías, Instituto de Agroquímica y Tecnología de Alimentos (IATA), Consejo Superior de Investigaciones Científicas (CSIC), Paterna, España.

Jorge A. Ramos-Hernández, Tecnológico Nacional de México, Instituto Tecnológico de Tepic, Nayarit, México.

Lorena Reyes-Vaquero, Centro de Investigación y Asistencia en Tecnología y Diseño del Estado de Jalisco A. C., Subsede Sureste, Mérida, Yucatán, Mexico.

Elizabeth Rubio-Rodríguez, Instituto Politécnico Nacional, Centro de Desarrollo de Productos Bióticos, Morelos, Mexico.

José Luis Trejo-Espino, Instituto Politécnico Nacional, Centro de Desarrollo de Productos Bióticos, Morelos, Mexico.

Gabriela Trejo-Tapia, Instituto Politécnico Nacional, Centro de Desarrollo de Productos Bióticos, Morelos, Mexico.

Pablo Emilio Vanegas-Espinoza, Instituto Politécnico Nacional, Centro de Desarrollo de Productos Bióticos, Morelos, Mexico.

Rosa Isela Ventura-Aguilar, CONACYT - Instituto Politécnico Nacional, Centro de Desarrollo de Productos Bióticos, Morelos, Mexico.

Daniel Arturo Zavala-Ortiz, Tecnológico Nacional de México, Instituto Tecnológico de Veracruz, Mexico.

Series Preface

BIOTECHNOLOGY – OUTSTANDING FACTS

The beginning of agriculture started about 12,000 years ago and, ever since, has played a key role in food production. We look to the farmers to provide the food we need but at the same time, now more than ever, to farm in a manner compatible with the preservation of the essential natural resources of the earth. In addition to the remarkable positive effects that farming has had throughout history, several undesirable consequences have been generated. The diversity of plants and animal species that inhabit the earth is decreasing. Intensified crop production has had undesirable effects on the environment (i.e., chemical contamination of groundwater, soil erosion, exhaustion of water reserves). If we do not improve the efficiency of crop production in the short term, we are likely to destroy the very resource base on which this production relies. Thus, the role of the so-called sustainable agriculture in the developed and underdeveloped world, where farming practices are to be modified so that food production takes place in stable ecosystems, is expected to be of strategic importance in the future; but the future has already arrived.

Biotechnology of plants is a key player in these scenarios of the twenty-first century; nowadays, molecular biotechnology, in particular, is receiving increasing attention because it has the tools of innovation for the agriculture, food, chemical, and pharmaceutical industries. It provides the means to translate an understanding of and an ability to modify, plant development and reproduction into enhanced productivity of traditional and new products. Plant products, either seeds, fruits, plant components, or extracts, are being produced with better functional properties and longer shelf life; and they need to be assimilated into commercial agriculture to offer new options to small, and more than small, industries and finally to consumers. Within these strategies, it is imperative to select crops with larger proportions of edible parts as well, thus generating less waste; it is also imperative to consider the selection and development of a more environment-friendly agriculture.

The development of research innovations for products is progressing, but the constraints of relatively long time frames to reach the marketplace, intellectual property rights, uncertain profitability of the products, consumer acceptance, and even the caution and fear with which the public may view biotechnology are tempering the momentum of all but the most determined efforts. Nevertheless, it appears uncontestable that the food biotechnology of plants and microbials is and will emerge as a strategic component to provide food crops and other products required for human well-being.

FOOD BIOTECHNOLOGY AND ENGINEERING SERIES / OCTAVIO PAREDES-LOPEZ, PHD, SERIES EDITOR

The Food Biotechnology and Engineering Series aims at addressing a range of topics around the edge of food biotechnology and the food microbial world. In the case of foods, it includes molecular biology, genetic engineering, and metabolic aspects, science, chemistry, nutrition, medical foods, and health ingredients, and processing and engineering with traditional- and innovative-based approaches. Environmental aspects to produce green foods cannot be left aside. At a world level, there are foods and beverages produced by different types of microbial technologies to which this series will give attention. It will also consider genetic modifications of microbial cells to produce nutraceutical ingredients, and advances in investigations on the human microbiome as related to diets.

TITLE: ADVANCES IN PLANT BIOTECHNOLOGY: *IN VITRO* PRODUCTION OF SECONDARY METABOLITES OF INDUSTRIAL INTEREST

Editors: Alma Angélica Del Villar-Martínez, Juan Arturo Ragazzo-Sánchez, Pablo Emilio Vargas-Espinoza, and Octavio Paredes-López

This book contains nine chapters written by 30 authors, 18 ladies and 12 gentlemen, from 12 different universities, centers of research, and technological and industrial organizations located in various countries. These researchers are doing basic and applied research, and innovation in several different fields of plant biotechnology.

As the editors of this book indicate, plant biotechnology produces a wide range of products and ingredients used in the agricultural, pharmaceutical, cosmetic, and industrial sectors in general. The use of plants as a rich source of secondary metabolites dates back to ancient times; the development of technology based on plant cell, tissue, and organ cultures has played a key role in the availability of metabolites, in conjunction with novel techniques for their extraction, encapsulation, handling, and application. Recent times have shown that the generation of new tools, such as metabolomics and other omics sciences, have also been fundamental to identify, characterize, and quantify the biochemical molecules involved in the structure and function of living organisms; possibilities for enhancing metabolite biosynthesis, under optimized conditions, are other important potential options. The necessity of finding solutions to nutraceutical global health and several industrial requirements, among other aspects, are doubtless of paramount importance as well.

In brief, the clear advantage of this publication is that each chapter is written by experts in the corresponding field, with the purpose of introducing the reader to the exciting world of plant biotechnology in general and its secondary metabolites to meet the challenges of the described concerns, and to the strategies of the human societies of the twenty-first century.

It is a good opportunity to express our full recognition and appreciation to all editors and authors for their excellent contributions, and to the editorial staff of CRC Press, especially to Ms. Laura Piedrahita and Mr. Stephen Zollo.

Preface

In its different fields, biotechnology constitutes an important discipline which plays a main role in our everyday life. In numerous situations, product and ingredient outcomes of biotechnological research are widely used in the agri-food, pharmaceutical, and industrial sectors. It is evident that the average population uses a wide range of chemical compounds in their daily activities, without realizing that their biological properties come mainly from plants and the microbial world.

Plants, as higher organisms, carry out diverse metabolic pathways, where the biosynthesis of metabolites occurs and which have been historically used in food, pharmacological, and cosmetic industries, among others. The interest in the use of plants as a source of useful metabolites for the treatment of diseases dates back to ancient times, from which traditional medicine arises.

It is pertinent to mention that the development of biotechnological techniques plays a key role in view of their remarkable characteristics and advantages; this is the case of plant cell, tissue, and organ culture. The importance of plant secondary metabolites in human health and in nutraceutical aspects, in conjunction with various biotechnological tools currently available nowadays, were the main reason for the generation of the present book entitled "Advances in Plant Biotechnology – *In vitro* Production of Secondary Metabolites of Industrial Interest".

This publication aims to integrate the results recently reported on plant cell, tissue, and organ culture; the use of medicinal plants; the study of plant secondary metabolites and their extraction techniques; the development and application of tools such as metabolomics, genetic transformation by *Agrobacterium*, and encapsulation of secondary metabolites; and the effect of some chemical compounds on human health.

Biotechnological tools, such as plant cell, tissue, and organ culture, have been fundamental for the scientific advances of the *in vitro* culture to produce secondary metabolites, and to enhance their biosynthesis, under controlled and optimized conditions. The use of hairy root cultures, generated by *Agrobacterium rhizogenes*, to synthesize chemical compounds of pharma-nutraceutical interest, taking advantage of their genetic stability and rapid growth, is also included. Strategies for obtaining metabolites of high industrial potential through *in vitro* culture are also described. In this sense, the most recent research advances, on the *in vitro* production of secondary metabolites with anti-inflammatory effects, is another outstanding contribution.

The state of the art of metabolomics management, extraction methods, and the encapsulation of plant-derived secondary metabolites are addressed as well, and the very beginning of the design for effective human application and release of chemical compounds inside the organism, with potential health benefits, and their remarkable roles and limitations are also analyzed. It should be mentioned that, in different sections of this publication, the use of the most recent omics and biotechnological strategies appear in a highly descriptive way.

For us, editors and authors, this project is a truly exciting adventure, where many concepts have been integrated to guide the reader in the generation of new ideas toward the development of products and ingredients that may contribute to solving global health and nutraceutical requirements. We would like to encourage undergraduate and graduate students, teachers and research professionals at college and university level, technical staff of processing industries, and private and public organizations involved with plant biotechnology in its different fields to share the reading of this book, which could be an explosive adventure of information. At the same time, we hope that it will be received with the same excitement with which it was conceived and developed.

CHAPTER 1

Cell Culture, Somatic Embryogenesis, and Plant Regeneration

Alma Angélica Del Villar-Martínez,
Yessica Casales Tlatilpa, and
Octavio Paredes-López

CONTENTS

1.1 INTRODUCTION

Plant tissue culture is an interesting tool for developing innovative technologies and obtaining important chemical compounds of value to human health, and it can be applied in research areas such as food, pharmacy, and agronomy, among others. Plant cell, tissue, and organ cultures can be regarded as an alternative strategy by which to conserve valuable plants and propagate those of commercial interest (Nowakowska et al. 2020).

Concerning the establishment of *in vitro* plant cell and organ cultures, some important stages will now be outlined.

DOI: 10.1201/9781003166535-1

1) Selection of plant material and establishment of conditions for aseptic cultures. It is necessary to select a young tissue because it may be less contaminated and is more susceptible to cell dedifferentiation. During the tissue disinfection process, it is common to use detergents, alcohol, water, and some biocidal solutions; eventually, antimicrobial agents are used to eliminate bacteria and fungi (Dar et al. 2022).

2) Development of dedifferentiated cells using plant growth regulators. At this stage, the culture medium and an appropriate combination of plant growth regulators must be selected for the cell dedifferentiation process. It is necessary to consider that each plant's response is different, and the selection of optimum plant growth regulator combinations is specific for each one.

3) Superiority over traditional propagation techniques. The *in vitro* culture techniques can include *in vitro* cell line establishment, *Agrobacterium*-mediated gene transfer, *in vitro* cell suspension culture, bioreactor culture, *in vitro* "hairy roots" production, and the use of genetically modified plant cultures which could be maintained for a long time, among others (De Schutter et al. 2022). The use of this technology represents important advantages over traditional methods because the establishment of *in vitro* cell, tissue, or organ cultures is achieved in a short time, unlike propagation under field or greenhouse conditions that take considerably longer because of environmental factors.

4) *In vitro* culture has no seasonal constraints but it is necessary to establish culture conditions and culture medium composition in which the biosynthesis of metabolites could be induced at any time of the year. In the same way, the combination of plant growth regulators is extremely important. All environmental factors provide signals recognized by cells and which are processed through metabolism; plants grown under greenhouse or field conditions are exposed to different factors, whereas *in vitro* cultures are developed under controlled environmental conditions which is an interesting option.

5) The production of disease-free plantlets. With *in vitro* plant culture, it is possible to obtain plants free of contamination (Wang et al. 2018). Aseptic plant cultures can be transported to other countries or different regions in the same country without the risk of carrying diseases, thus avoiding contamination, and promoting germplasm

conservation. Moreover, plants propagated by *in vitro* culture are genetically identical to the original plant and it is possible to maintain a bank of plants that can be used as explant sources, regardless of climatic conditions (George et al. 2022, Sharma et al. 2022).

6) Induction of biosynthesis or accumulation of specific plant compounds. The selection of specific media for the study of medicinal plants is decisive. Plant cells respond to the stimuli received so that metabolite biosynthesis pathways can be stimulated or blocked. At this stage, it is possible to make use of modifications in the culture medium components or to apply stimuli, such as storage temperature, pH, hydric or saline stress, or chemical agents to act as elicitors or precursors, all of which may modify the biosynthetic pathways (Abdelazeez et al. 2022).

1.2 PLANT CELL AND TISSUE CULTURE FOR PLANT MICROPROPAGATION

In tissue culture techniques, under aseptic conditions, manipulation of tissue or organ cells results in the growth of undifferentiated cells in a culture medium under suitable conditions, in which all environmental (light, temperature, and humidity) and culture medium (nutrients, etc.) factors are under control.

Plant cell, tissue, and organ cultures can be regarded as suitable alternative approaches to conserve many valuable plants and achieve their propagation, since some crops suffer from significant difficulties due to poor seed quality, low germination rate, or low seedling survival under natural field conditions.

Somatic embryogenesis (SE) is a remarkable process by which embryos can be produced from somatic cells. Somatic embryos go through morphological stages similar to zygotic embryos, described in dicotyledonous angiosperms as globular, heart, torpedo, and cotyledon stages (Rose and Song 2017). Two different types of somatic embryogenic routes are generally involved in plants, direct and indirect.

With direct SE, there is no dedifferentiation stage, and embryogenic cell formation can be completed directly from the surface of explants, in which minimal genetic reprogramming is involved; indirect somatic embryogenesis, on the other hand, is a multistep regeneration process including somatic embryo formation, maturation, and conversion that requires major reprogramming (Zhang et al. 2021).

Coriandrum sativum or coriander is one of the most popularly used herbs and spices in cooking worldwide, and its medicinal values have been recognized since ancient times. *C. sativum* contains bioactive phytochemicals that account for a wide range of biological activities including antioxidant, anticancer, neuroprotective, anxiolytic, anticonvulsant, analgesic, migraine-relieving, hypolipidemic, hypoglycemic, hypotensive, antimicrobial, and anti-inflammatory activities. The major compound, linalool, abundantly present in coriander seeds, is noted for its abilities to modulate many key pathogenesis pathways of human diseases (Prachayasittikul et al. 2018). Figure 1.1 shows embryogenic processes in the *in vitro* culture of coriander. Figure 1.1A shows different stages in the process to achieve complete dedifferentiation of hypocotyls in Murashige & Skoog (MS) medium (Murashige and Skoog 1962) supplemented with 2,4-dichlorophenoxyacetic acid (2,4-D, 1 mg/L). The hypocotyl explant of coriander plants began dedifferentiation at seven days of culture; the formation of callus occurred at the ends of the explant and an increase in the size of the explant was observed between 15 and 20 days of culture because of the absorption of water, and the tissue was completely dedifferentiated and highly dispersible at 25 days of culture. Calli were subcultured, and they changed to globular structures at 15 days of culture, which have been termed pro-embryogenic calli.

It was previously mentioned that plant tissues are sensitive to changes in the content of nutrients in the culture medium. Figure 1.1B and C shows the callus response to a concentration gradient (30, 60, 75, and 90 g/L) of glucose or sucrose, and the effect on germination of pro-embryogenic calli from coriander, due to changes in the carbohydrate source. The highest concentration of a specific sugar (90 g/L) resulted in five or even fewer somatic embryos being obtained per explant and the tissue turning brown. Medium that contained 60 g/L of carbon source showed better results, with sucrose at 60 g/L gave a better response than glucose (Figure 1.1C), which exhibited abundant embryogenesis, and the tissue appearance was acceptable in MS medium without plant growth regulators. *C. sativum* is a good example to demonstrate the response of the tissue to changes in the composition of the culture medium, which induces embryo development. Finally, the four characteristic stages of the embryogenesis process were identified: globular, torpedo, heart, and cotyledon stages, as shown in Figure 1.1D.

Tissue culture and its importance in the regeneration of coffee plants is described in detail at cellular level by Ferrari et al. (2021), where the

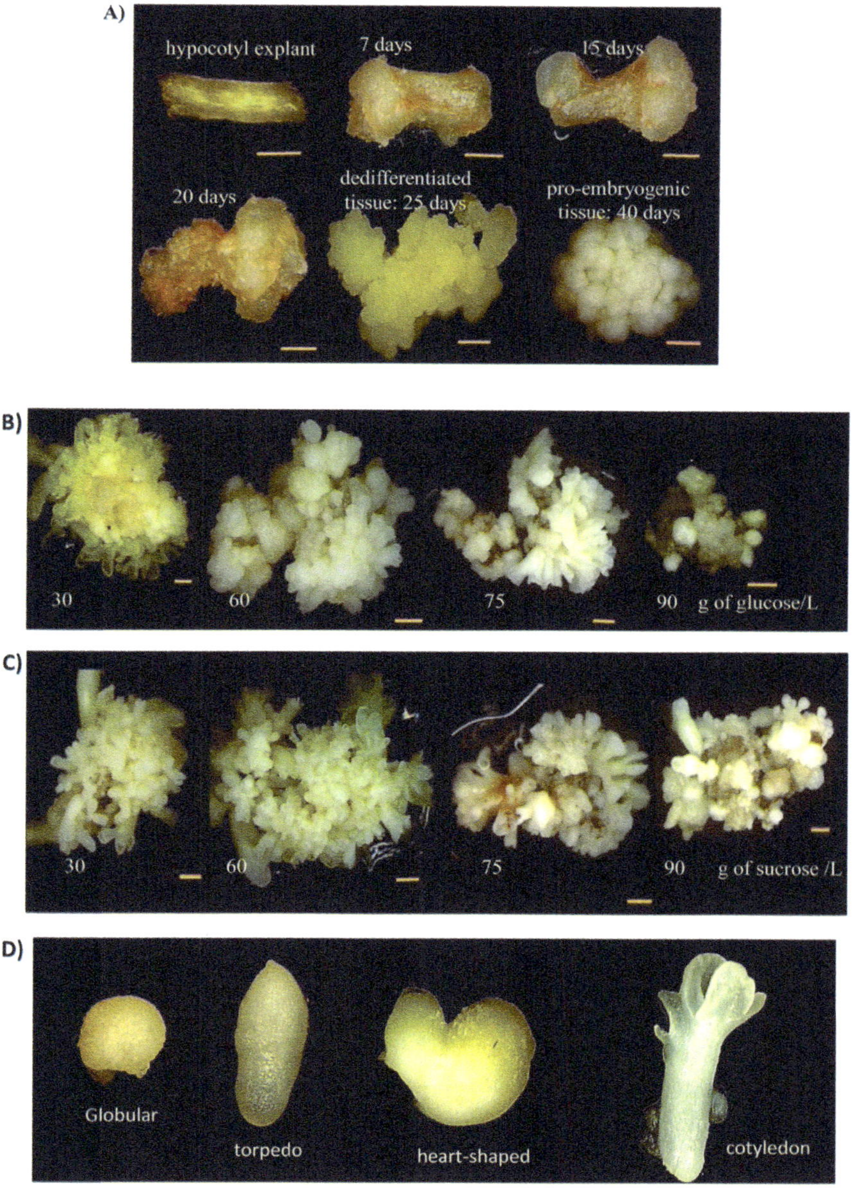

FIGURE 1.1 Embryogenic processes in the *in vitro* culture of coriander (*Coriandrum sativum*) in MS medium supplemented with 2,4-D (1 mg/L). (A) Dedifferentiation process of hypocotyl explant (B) show the development of embryos in MS medium with glucose (B) or (C) sucrose, (D) stages of embryogenesis process. Bar= 1 cm.

embryogenic process is reported. Plants of *Coffea arabica* cultivar Mundo Novo were collected. This is the most-widely cultivated coffee cultivar in Brazil, with a high capacity to adapt to different regions; it has excellent agronomic characteristics such as large plant size, red fruits, medium ripeness, and excellent drink quality. Ferrari et al. (2021) present the basic description of direct and indirect somatic embryogenesis in coffee.

1) Somatic embryogenesis (SE) involves cell totipotency, whereby somatic cells from plant tissues undergo reprogramming to generate embryonic cells, through a series of developmental stages like those occurring in zygotic embryogenesis.
2) A pro-embryogenic mass was identified in DSE (direct somatic embryogenesis) and ISE (indirect somatic embryogenesis) and originated from mitotically active cells present in the spongy parenchyma.
3) The development of the pro-embryogenic mass from callus by DSE and ISE occurred in the initial phase of SE, which demonstrated a correlation between the presence of vascular tissue and the early formation of these structures.
4) *In vitro*, the embryos obtained via the ISE pathway grew faster than the embryos obtained via the DSE pathway.

Beigmohamadi et al. (2019) reported the propagation of *Plumbago europaea* because of its capacity to produce large amounts of plumbagin as a bioactive naphthoquinone, which has been reported to possess antimicrobial, anticancer, antifertility, pesticidal, and antimalarial activities. In this study, an efficient method for *in vitro* regeneration of *P. europaea* was developed by direct and indirect organogenesis.

Spontaneous regeneration of plantlets derived from hairy root cultures has been reported by Vargas-Morales et al. (2022). In this work, the regeneration process in two hairy root lines was obtained through indirect somatic embryogenesis. Moreover, histological analysis of the hairy-root-derived callus samples showed the development of somatic embryos.

1.3 IMPORTANCE OF PLANT CELL, TISSUE, AND ORGAN CULTURE

The interest in medicinal plants arises because, since ancient times, they have been used to treat different human diseases, demonstrating biological activities such as anti-inflammatory, anxiolytic, antipyretic, antiviral, and antibacterial activities, among others, because of a diversity of chemical compounds called secondary metabolites.

Secondary metabolites are small molecules manufactured by plants that make them resilient to their environment and exert a wide range of effects on the plant itself or other living organisms (Teoh 2016). The metabolites synthesized by the plant during secondary metabolism are very important, due to the different benefits observed concerning human health. Some widely studied groups of metabolites are terpenes, nitrogen-containing secondary metabolites, and phenolic compounds, all of which are derived initially from primary metabolism (Figure 1.2).

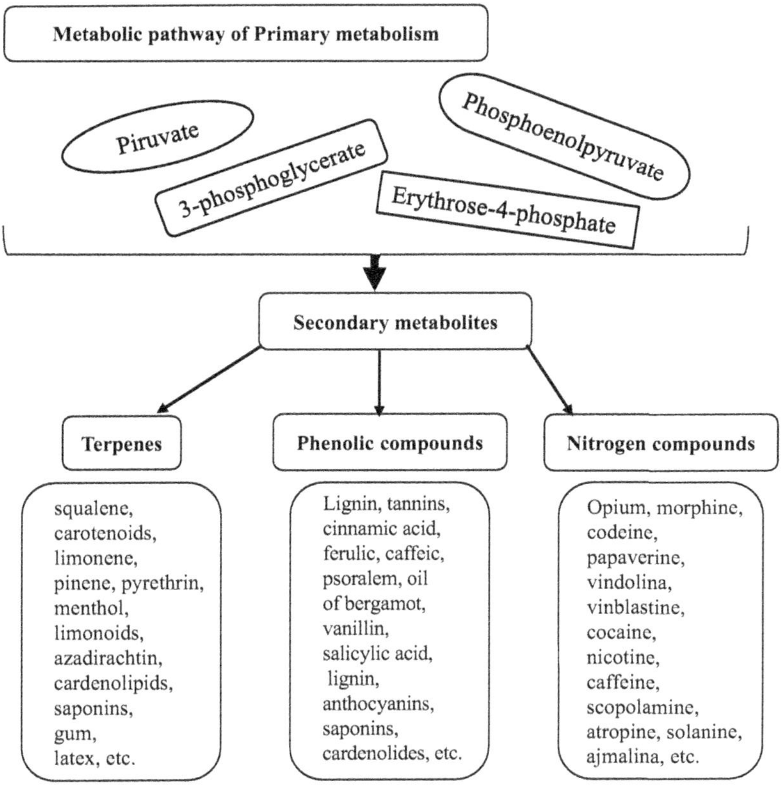

FIGURE 1.2 Some groups of metabolites well studied in plant cell tissue culture: terpenes, nitrogen-based secondary metabolites, and phenolic compounds, all of which are derived from primary metabolite precursors.

Recent evidence indicates that the benefits arising from the consumption of those plant compounds may be because of their antioxidant, anti-inflammatory, antitumor, or anticancer properties, among others (Ochoa-Jiménez et al. 2019).

Secondary metabolites are synthesized and stored in different plant tissues and organs; their biosynthesis and accumulation depend on environmental conditions and development of the plants, so activation or inhibition of secondary metabolite production occurs when environmental conditions affect the biosynthesis pathway.

Food and pharmaceutical ingredients like sweeteners, essential oils, colorants, nutraceuticals, flavors, and antioxidants have been produced by plant cell culture technology.

In brief, plant cell, tissue, and organ culture have several advantages over traditional plant culture and it is possible to achieve: mass propagation of plants, especially for species that are difficult to propagate by other methods, micropropagation of endangered species, propagation of plants in considerably reduced time, cloning of individuals with highly desirable agronomic characteristics throughout the year, production of virus-free plants and synthetic seeds for germplasm conservation, production of secondary metabolites by modifying metabolite biosynthetic pathways, production of new hybrids, genetic improvement of plants, including transgenic plants, seed germination, and several physiological studies, among others.

Mass production by cell culture is of special importance because *in vitro* plant culture, for which there are no seasonal or regional restrictions, may be used as an alternative to traditional plant culture; to produce secondary metabolites of commercial interest, mass plant micropropagation or genetically modified plants, cells, organs, under controlled settings.

Atropa acuminata is a critically endangered annual herb growing in Kashmir Himalaya and is used as a source of medicine for many ailments like arthritis, muscular pain, and joint pain. Given the medicinal importance of *Atropa acuminata*, the plant needs a tissue culture protocol for mass propagation of this rare species. Dar et al. (2022) reported a highly reliable and reproducible tissue culture protocol for this valuable plant, as well as evaluation of biochemical attributes and antioxidant activity. *In vitro*-grown plants provide an important alternative source of primary and secondary metabolites, and can provide some natural biochemical and antioxidant products that can be used in the pharmaceutical industry.

1.4 MEDICINAL PLANTS AND *IN VITRO* CELL CULTURE

The technology of *in vitro* plant culture offers several advantages due to its characteristics. Medicinal plants have enormous practical potential which deserves to be used in a rational and intelligent way. *In vitro* culture eliminates seasonal and environmental limitations, since it is possible to obtain enriched extracts of specific compounds as a result of cultures which can be activated toward the accumulation of specific metabolites through strategies such as the use of elicitors, changes in the composition of the culture medium, changes of environmental conditions, or even genetic modifications (Table 1.1).

Siham et al. (2021) conducted a literature review of studies where different types of secondary metabolites from medicinal plants, with a variety of uses, such as antioxidant, immunomodulatory, antidepressant, and anti-inflammatory, were produced by several plant tissue culture techniques. Abdelazeez et al. (2022) reported for the first time the production of tropane alkaloids by cell suspension cultures from root explants of *Hyoscyamus muticus*. Results showed that the plant cell culture of *H. muticus* could be an excellent alternative to whole plant culture for producing atropine alkaloids; *in vitro* cultures could synthesize the tropane alkaloids in quantities greater than in the original intact plant. *H. muticus* is a valuable medicinal plant because of the synthesis of alkaloids such as scopolamine, hyoscyamine, and atropine, which have a wide range of pharmacological actions. On the other hand, *Rosmarinus officinalis* (Lamiaceae), rosemary, is an aromatic plant popular worldwide, mainly due to its uses in traditional medicine as an anti-inflammatory, diuretic, and antimicrobial agent, as well as in the prevention and treatment of other diseases, and as a herb in cookery. These biological activities are mainly related to the presence of phenolic and terpenoid compounds; Perez-Mendoza et al. (2019) carried out a culture of plant cells and the results showed that the dedifferentiated cells biosynthesized chemical compounds found in highly specialized organs such as leaves. Urquiza-López et al. (2021) reported the metabolite profiling of rosemary cell lines and their antiproliferative potential against cancer cell lines. Three stable rosemary cell lines, named *Ro*G, *Ro*Y, and *Ro*W, were obtained and characterized. The chemical profile of each cell line was analyzed and it was observed that the lower levels of rosmarinic acid found in the *Ro*W extract compared with its higher content of humulone derivatives, rosmaridiphenol, and hinokione isomer, were associated with the observed greater activity of *Ro*W against HT-29 colon cancer cell

TABLE 1.1 Secondary Metabolites Produced by Plant Cell and Organ Culture

Species	Culture	Metabolites	Reference
Melia azedarach	Callus	Quercetin, rutin and kaempferol	Ahmadpoor et al. 2022
Calendula officinalis	Hairy roots	Steroids and triterpenoids	Rogowska et al. 2022
Taxus cuspidata	Cell suspension	Paclitaxel	Yamamoto et al. 2022
Mucuna imbricata	Callus	L-DOPA	Suryawanshi et al. 2022
Taxus× media	Hairy roots	Paclitaxel	Sykłowska-Baranek et al.. 2022
Rubus chamaemorus, Rubus saxatilis, Vaccinium vitis-idaea, Rubus arcticus Fragaria × ananassa	Callus	Phenolic compounds and carotenoids	Rischer et al. 2022
Curcuma amada	Micropropagated plants	Antioxidants	Behera et al. 2022
Ocimum sp.	Hairy roots	Terpenes	Pandey et al. 2022
Gazania rigens	Cell suspension	Terpenes	Mahood et al. 2022
Ocimum basilicum	Hairy roots	Secondary metabolites	Sathasivam et al. 2022
Gynostemma pentaphyllum	Cell suspension	Saponins	Quang et al. 2022
Solanum lycopersicum cv. Micro-Tom	Callus	Antioxidants	Gogliettino et al. 2022
Salvia dominica	Hairy roots	Proteins	Boccia et al. 2022
Hyoscyamus muticus	Cell suspension	Atropine	Abdelmaksood et al. 2022
Aralia elata	Cell suspension	Oleanolic acid and flavonoids	Sui et al. 2022
Atropa acuminata	Regenerated plants	Phenolic compounds and flavonoids	Dar et al. 2022
Rosmarinus officinalis	Callus	Flavonoids and terpenes	Urquiza-López et al. 2021
Ballota nigra	Regenerated plants	Phenolic compounds and flavonoids	Younessi-Hamzekhanlu et al. 2021
Hyssopus officinalis	Callus	Polyphenolic compounds	Babich et al. 2021

(*Continued*)

TABLE 1.1 (CONTINUED) Secondary Metabolites Produced by Plant Cell and Organ Culture

Species	Culture	Metabolites	Reference
Vaccinium corymbosum	Callus	Anthocyanins	El-Dis et al. 2021
Brassica oleracea	Hairy roots	Glucoraphanin and sulforaphane	Zhang et al. 2021
Ballota nigra	Regenerated plants	Phenolic compounds and flavonoids	Younessi-Hamzekhanlu et al. 2021
Rosmarinus officinalis	Callus	Phenolic compounds and terpenes	Pérez-Mendoza et al. 2020
Ocimum basilicum	Cell suspension	Triterpenoids	Mehring et al. 2020
Jatropha curcas	Callus	Phorbol esters and flavonoids	Leyva-Padrón et al. 2020
Ophiorrhiza pumila	Hairy roots	Camptothecin	Shi et al. 2020
Tripterygium wilfordii	Hairy roots	Wilforgine and wilforine	Zhang et al. 2020
Ardisia crenata	Hairy roots	Ardicrenin	Hu et al. 2020
Vaccinium arctostaphylos	Regenerated plants	Phenolic compounds and flavonoids	Bakhshipour et al. 2019
Achillea gypsicola	Cell suspension	Camphor and phenolic compounds	Muhammed et al. 2019
Lycium ruthenicum	Hairy roots	Kukoamine A	Chahel et al. 2019
Plumbago europaea	Cell suspension	Plumbagin	Beigmohamadi et al. 2019

lines. It is also interesting to report that Stefanowicz-Hajduk et al. (2021) analyzed the phytochemical composition of *Kalanchoe daigremontiana* water leaves and root plant extracts and estimated its cytotoxic activity against human ovarian cancer SKOV-3 cells.

1.5 BIOACTIVE COMPOUNDS AND THE ENCAPSULATION PROCESS

Bioactive compounds often have low water solubility, and they can also be unstable in the presence of oxygen, light, heat, and humidity. In addition, the functionality of bioactive compounds is inherently related to their bioavailability; they must be effectively absorbed in the gut, then carried out into the circulatory system, before reaching the target cells. Moreover, these compounds can be degraded due to pH changes (stomach/small intestine). In this sense, stabilization technologies, such as encapsulation and adsorption, could improve the stability and bioavailability of the bioactive compounds (Koop et al. 2022). Encapsulation is described as a method of entrapment of a core material, such as an active ingredient, within another solid or liquid immiscible substance, thereby producing capsules with diameters ranging in general from 10 nm to 10 μm. Core materials include solid particles, liquid droplets, or gas bubbles. The immiscible substance is also known as the carrier(s) or wall material, shell, coating, membrane, external phase, or matrix. Generally, one can differentiate between two main forms and structures (morphologies) of encapsulated systems, namely core-shell type (capsules) and matrix (spheres) type (Vincekovic et al. 2017). Several techniques can be employed for microencapsulation, such as freeze drying, spray drying, coacervation, emulsion-based processes, co-crystallization, extrusion, supercritical fluid-base processes, spray chilling, ultrasound-based processes, and liposomes. Each of these methods has its specific strengths and weaknesses in terms of encapsulation, protection, delivery, cost, regulatory status, ease of use, biodegradability, and biocompatibility (Chawda et al. 2017). An efficient system for encapsulating bioactive compounds must have the following properties: 1) it produces food-grade products; 2) it incorporates bioactive compounds in food matrices with physiochemical stability and minimal impact on the product's sensory properties; 3) it protects encapsulated compounds from interactions with other ingredients and degradation due to temperature, light, or pH; 4) it maximizes the absorption of encapsulated compounds after consumption, ensuring controlled release in response to a specific stimulus; and 5) exhibits scaling-up potential for industrial production (Mandaji et al. 2022). Among the encapsulation techniques, spray drying is simple, fast, and feasible to scale up. Thus, it has been widely used in the chemical, food, and pharmaceutical industries for the encapsulation of valuable active ingredients. Spray-drying equipment is commercially available worldwide, while the dehydrated products obtained exhibit valuable qualities, such as low degradation rates and high stability properties (Piñón-Balderrama et al. 2020).

The microencapsulation process by spray drying comprises the following main steps: 1) the "feed" preparation, such as a tissue culture extract, that may be a solution, a suspension, or an emulsion, coats the wall and core materials; 2) atomization of the feed into a small droplet; 3) rapid drying of liquid droplets in contact with a stream of hot gas, resulting in the instantaneous formation of the microparticles in the form of powder; and finally, 4) powder recovery. The temperature of the drying inlet air is usually between 110 and 220°C, and the exposure time of the feed solution to these high temperatures is only a few milliseconds. The temperature inside the microparticles, where the core material is present, is normally below 80°C, which helps to minimize the thermal degradation of the bioactive material (Corrêa-Filho et al. 2019).

A range of natural or synthetic polymers that meet the safety requirements of governmental agencies, such as the US Food and Drug Administration (FDA) or the European Food Safety Authority (EFSA), among others, can be used to encapsulate food and herbal products. Polymers, such as polysaccharides, maltodextrin, gum arabic, modified starches, and chitosan, as well as mixtures of them, are frequently selected as coating materials (Mudalip et al. 2021). Alvarado-Palacios et al. (2015) evaluated nanoencapsulated *K. daigremontiana* extracts based on their pharmacological properties; *in vitro* experiments showed that nanocapsules that contained an aquoethanolic extract had a greater cytotoxic effect on the MDA-MB231 metastatic breast cancer cell line than the non-encapsulated aquoethanolic extract. Additionally, when the cytotoxic effect was compared in the non-cancerous breast cell line MCF 10A, the authors found no cytotoxic effect of the nanocapsules containing the aquoethanolic extract of *K. daigremontiana*, whereas the aquoethanolic extract, which was not encapsulated, had a significant cytotoxic effect. The results could be attributed to the small size of the capsules that could be introduced into tumor cells but not into normal breast cells, due to the difference in surface morphology of the cells. Moreover, the size of nanocapsules would help the penetration of nanoparticles into the cell and thereby carry out the combined processes of erosion and diffusion of the polymeric matrix, releasing the active ingredient of *K. daigremontiana* from the nanocapsules within, causing cell death; this contribution is very important for cancer treatment.

1.6 CONCLUSIONS

Plant cell culture offers the possibility of developing biotechnological processes to obtain compounds of commercial interest. The adjustment of the culture conditions offers the opportunity to obtain important chemical compounds in less time compared with medicinal plants grown in the field. The methodologies addressed by cell and tissue cultures can be used to improve crops; obtain metabolites of industrial and economic interest in less time; and generate plants that are difficult to obtain by traditional means under controlled conditions. Obtaining specific cell lines, or even genetically modified organisms, with outstanding potential to produce compounds useful in medicine and for industrial purposes for the well-being of society, represent key tools for the advancement of basic and applied research in conjunction with innovative technologies. The goal of this chapter is to support the aims herewith described.

ACKNOWLEDGMENTS

AADVM acknowledges the financial support provided by SIP Instituto Politécnico Nacional Mexico (IPN/SIP 20211235 and 20220810), and deeply grateful to Iain André Del Villar Martínez for the great support, OPL recognizes the membership distinction for life by El Colegio de Sinaloa (Mexico).

REFERENCES

Abdelazeez W. M. A., K. Y. Anatolievna, K. L. Zavdetovna, A. G. Damirovna, and G. R. A. El-Dis. 2022. "Enhanced productivity of atropine in cell suspension culture of *Hyoscyamus muticus* L." *In Vitro Cellular and Developmental Biology Plant* 58: 593–605. https://doi.org/10.1007/s11627-022-10262-z

Ahmadpoor F., N. Zare, R. Asghari, and P. Sheikhzadeh. 2022. "Sterilization protocols and the effect of plant growth regulators on callus induction and secondary metabolites production in in vitro cultures *Melia azedarach* L." *AMB Express* 12(1): 1–12. https://doi.org/10.1186/s13568-022-01343-8

Alvarado-Palacios Q. G., E. N. San Martin-Martinez, C. Gomez-García, C. C. Estanislao-Gomez, and R. Casañas-Pimentel. 2015. "Nanoencapsulation of the Aranto (*Kalanchoe daigremontiana*) aquoethanolic extract by nanospray dryer and its selective effect on breast cancer cell line." *International Journal of Pharmacognosy and Phytochemical Research* 7(5): 888–895.

Babich O., S. Sukhikh, A. Pungin, L.Astahova, E. Chupakhin, D. Belova, A. Prosekov, and S. Ivanova. 2021. "Evaluation of the conditions for the cultivation of callus cultures of *Hyssopus officinalis* regarding the yield of polyphenolic compounds." *Plants* 10: 915. https://doi.org/10.3390/plants10050915

Bakhshipour M., M. Mafakheri, M. Kordrostami, A. Zakir, N. Rahimi, F. Feizi, and M. Mohseni. 2019. "*In vitro* multiplication, genetic fidelity and phytochemical potentials of *Vaccinium arctostaphylos* L.: An endangered medicinal plant." *Industrial Crops and Products* 141: 111812. https://doi.org/10.1016/j.indcrop.2019.111812

Behera S., M. Kumari, R. K. Meher, S. Mohapatra, S. K. Madkami, P. K. Das, P. K. Naik, and S. K. Naik. 2022. "Phytochemical fidelity and therapeutic activity of micropropagated *Curcuma amada* Roxb.: A valuable medicinal herb." *Industrial Crops and Products* 176: 114401. https://doi.org/10.1016/j.indcrop.2021.114401.

Beigmohamadi M., A. Movafeghi, A. Sharafi, S. Jafari, and H. Danafar. 2019. "Cell suspension culture of *Plumbago europaea* L. towards production of plumbagin." *Iranian Journal of Biotechnology* 20: e2169. https://doi.org/10.21859/ijb.2169

Boccia E., M. Alfieri, R. Belvedere, V. Santoro, M. Colella, P. Del Gaudio, M. Moros, F. Dal Piaz, A. Petrella, A. Leone, and A. Ambrosone. 2022. "Plant hairy roots for the production of extracellular vesicles with antitumor bioactivity." *Communications Biology* 5: 848. https://doi.org/10.1038/s42003-022-03781-3

Chahel A. A., S. Zeng, Z. Yousaf, Y. Liao, Z. Yang, X. Wei, and W. Ying. 2019. "Plant-specific transcription factor LrTCP4 enhances secondary metabolite biosynthesis in *Lycium ruthenicum* hairy roots." *Plant Cell Tissue and Organ Culture* 136: 323–337.

Chawda P. J., J. Shi, S. Xue, and S. Y. Quek. 2017. "Co-encapsulation of bioactives for food applications." *Food Quality and Safety* 1(4): 302–309. https://doi.org/10.1093/fqsafe/fyx028

Corrêa-Filho L. C., M. Moldão-Martins, and V. D. Alves. 2019. "Advances in the application of microcapsules as carriers of functional compounds for food products." *Applied Sciences* 9(3): 571. https://doi.org/10.3390/app9030571

Dar S. A., I. A. Nawchoo, S. Tyub, and A. N. Kamili. 2022. "*In vitro* culture and biochemical and antioxidant potential of the critically endangered medicinal plant *Atropa acuminata* Royle ex Lindl of Kashmir Himalaya." *In Vitro Cellular and Developmental Biology-Plant* 58: 540–550. https://doi.org/10.1007/s11627-022-10271-y

De Schutter K., I. Verbeke, D. Kontogiannatos, P. Dubruel, L. Swevers, E. J. M. Van Damme, and G. Smagghe. 2022. "Use of cell cultures in vitro to assess the uptake of long dsRNA in plant cells." *In Vitro Cellular and Developmental Biology - Plant* 58: 511–520. https://doi.org/10.1007/s11627-022-10260-1

El-Dis G. R. A., K. L. Zavdetovna, A. A. Nikolaevich, W. M. A. Abdelazeezand, and T. O. Arnoldovna. 2021. "Influence of light on the accumulation of anthocyanins in callus culture of *Vaccinium corymbosum* L. cv. Sunt Blue Giant." *Journal of Photochemistry and Photobiology* 8: 100058. https://doi.org/10.1016/j.jpap.2021.100058

Ferrari F. I., G. A. Marques, W. L. Sachetti, B. B. Biazotti, M. P. Passos, J. A. S. de Almeida, J. M. C. Mondego, and J. L. Mayer. 2021. "Comparative ontogenesis of *Coffea arabica* L. somatic embryos reveals the efficiency of regeneration modulated by the explant source and the embryogenesis pathway." *In Vitro Cellular and Developmental Biology - Plant* 57: 796–810. https://doi.org/10.1007/s11627-021-10200-5

George N. M., S. B. Raghav, and D. Prasath. 2022 "Direct *in vitro* regeneration of medicinally important Indian and exotic red-colored ginger (*Zingiber officinale* Rosc.) and genetic fidelity assessment using ISSR and SSR markers." *In Vitro Cellular and Developmental Biology - Plant* 58: 551–558. https://doi.org/10.1007/s11627-022-10268-7

Gogliettino M., S. Arciello, F. Cillo, A. V. Carluccio, G. Palmieri, F. Apone, R. L. Ambrosio, A. Anastasio, L. Gratino, A. Carola, and E. Cocca. 2022. "Recombinant expression of archaeal superoxide dismutases in plant cell cultures: A sustainable solution with potential application in the food industry." *Antioxidants* 11: 1731. https://doi.org/10.3390/antiox11091731

Hu J., S. Liu, Q. Cheng, S. Pu, M. Mao, Y. Mu, F. Dan, J. Yang, and M. Ma. 2020. "Novel method for improving ardicrenin content in hairy roots of *Ardisia crenata* Sims plants." *Journal of Biotechnology* 311: 12–18. https://doi.org/10.1016/j.jbiotec.2020.02.009

Koop B. L., M. N. da Silva, F. D. da Silva, K. T. L. dos Santos, L. S. Soares, C. J. de Andrade, G. A. Valencia, and A. R. Monteiro. 2022. "Flavonoids, anthocyanins, betalains, curcumin, and carotenoids: Sources, classification and enhanced stabilization by encapsulation and adsorption." *Food Research International*, 110929. https://doi.org/10.1016/j.foodres.2021.110929

Leyva-Padrón G., P. E. Vanegas-Espinoza, S. Evangelista-Lozano, A. A. Del Villar-Martínez, and C. Bazaldúa. 2020. "Chemical analysis of callus extracts from toxic and non-toxic varieties of *Jatropha curcas* L." *Peer Journal* 8: e10172. https://doi.org/10.7717/peerj.10172

Mahood H. E., V. Sarropoulou, and T. T. Tzatzani. 2022. "Effect of explant type (leaf, stem) and 2,4-D concentration on callus induction: Influence of elicitor type (biotic, abiotic), elicitor concentration and elicitation time on biomass growth rate and costunolide biosynthesis in gazania (*Gazania rigens*) cell." *Bioresources and Bioprocessing* 9: 100. https://doi.org/10.1186/s40643-022-00588-2

Mandaji C. M., R. da Silva Pena, and R. Campos Chisté. 2022. "Encapsulation of bioactive compounds extracted from plants of genus Hibiscus: A review of selected techniques and applications." *Food Research International* 151: 110820. https://doi.org/10.1016/j.foodres.2021.110820

Mehring A., J. Haffelder, J. Chodorski, J. Stiefelmaier, D. Strieth, and R. Ulber. 2020. "Establishment and triterpenoid production of *Ocimum basilicum* cambial meristematic cells." *Plant Cell Tissue and Organ Culture* 143: 573–581. https://doi.org/10.1007/s11240-020-01942-y

Mudalip S. K. A., M. N. Khatiman, N. A. Hashim, R. C. Man, and Z. I. M. Arshad. 2021. "A short review on encapsulation of bioactive compounds using different drying techniques." *Materials Today: Proceedings* 42(1): 288–296. https://doi.org/10.1016/j.matpr.2021.01.543

Muhammed A A., M. K. Şevket, A. Ahmet, M. Ö. Mehmet, and B. A. Ebru. 2019. "Effects of methyl jasmonate and salicylic acid on the production of camphor and phenolic compounds in cell suspension culture of endemic

Turkish yarrow (*Achillea gypsicola*) species." *Turkish Journal of Agriculture and Forestry* 43: 3.

Murashige T., and F. Skoog. 1962. "A revised medium for rapid growth and bioassays with tobacco tissue cultures." *Physiologia Plantarum* 15(3): 473–497. https://doi.org/10.1111/j.1399-3054.1962.tb08052.x

Nowakowska M., Ž. Pavlović, M. Nowicki, S. L. Boggess, and R. N. Trigiano. 2020. "In vitro propagation of an endangered *Helianthus verticillatus* by axillary bud proliferation." *Plants* 9(6): 712. https://doi.org/10.3390/plants9060712

Ochoa-Jiménez V. A., J. C. Tafolla-Arellano, G. Berumen-Varela, and M. E. Tiznado-Hernández. 2019. "Biotechnology of horticultural commodities." In *Postharvest Technology of Perishable Horticultural Commodities*, edited by Elhadi M. Yahia, 695–708. Woodhead Publishing. https://doi.org/10.1016/B978-0-12-813276-0.00021-3

Pandey P., S. Singh, A. S. Negi, and S. Banerjee. 2022. "Harnessing the versatility of diverse pentacyclic triterpenoid synthesis through hairy root cultures of various *Ocimum* species: An unprecedented account with molecular probing and up-scaling access." *Industrial Crops and Products* 177: 114465. https://doi.org/10.1016/j.indcrop.2021.114465

Pérez-Mendoza M. B., L. Llorens-Escobar, P. E. Vanegas-Espinoza, A. Cifuentes, E. Ibáñez, and A. A. Del Villar-Martínez. 2019. "Chemical characterization of leaves and calli extracts of *Rosmarinus officinalis* by UHPLC-MS." *Electrophoresis* 41(20): 1776–1783. https://doi.org/10.1002/elps.201900152

Piñón-Balderrama C. I., C. Leyva-Porras, Y. Terán-Figueroa, V. Espinosa-Solís, C. Álvarez-Salas, and M. Z. Saavedra-Leos. 2020. "Encapsulation of active ingredients in food industry by spray-drying and nano spray-drying technologies." *Processes* 8(8): 889. https://doi.org/10.3390/pr8080889

Prachayasittikul V., S. Prachayasittikul, S. Ruchirawat, and V. Prachayasittikul. 2018. "Coriander (*Coriandrum sativum*): A promising functional food toward the well-being." *Food Research International* 105: 305–323. https://doi.org/10.1016/j.foodres.2017.11.019

Quang, H. T., P. T. D. Thi, D. N. Sang, T. T. N. Tram, N. D. Huy, and T. Q. Dung. 2022. "The Q.T.T. Effects of plant elicitors on growth and gypenosides biosynthesis in cell culture of giao co lam (*Gynostemma pentaphyllum*)." *Molecules* 27: 2972.

Rischer H., L. Nohynek, R. Puupponen-Pimiä, J. Aguiar, G. Rocchetti, L. Lucini, J. S. Câmara, T. M. Cruz, M. B. Marques, and D. Granato. 2022. "Plant cell cultures of Nordic berry species: Phenolic and carotenoid profiling and biological assessments." *Food Chemistry* 366: 130571. https://doi.org/10.1016/j.foodchem.2021.130571

Rogowska A., M. Stpiczyńska, C. Pączkowski, and A. Szakiel. 2022. "The influence of exogenous jasmonic acid on the biosynthesis of steroids and triterpenoids in *Calendula officinalis* plants and hairy root culture." *International Journal of Molecular Sciences* 23(20): 12173. https://doi.org/10.3390/ijms232012173

Rose R. J., and Y. Song. 2017. "Somatic Embryogenesis." In *Encyclopedia of Applied Plant Sciences*, edited by Brian Thomas, Brian G. Murray, and

Denis J. Murphy, 474–479. Academic Press. https://doi.org/10.1016/B978-0-12-394807-6.00147-7

Sathasivam R., M. Choi, R. Radhakrishnan, H. Kwon, J. Yoon, S. H. Yang, J. K. Kim, Y. S. Chung, and S. U. Park. 2022. "Effects of various *Agrobacterium rhizogenes* strains on hairy root induction and analyses of primary and secondary metabolites in *Ocimum basilicum*." *Frontiers in Plant Science* 13: 983776. https://doi.org/10.3389/fpls.2022.983776

Sharma N., E. V. Malhotra, R. Chandra, R. Gowthami, S. M. Sultan, S. Bansal, M. Shankar, and A. Agrawal. 2022. "Cryopreservation and genetic stability assessment of regenerants of the critically endangered medicinal plant *Dioscorea deltoidea* Wall. ex Griseb. for cryobanking of germplasm." *In Vitro Cellular and Developmental Biology - Plant* 58: 521–529. https://doi.org/10.1007/s11627-022-10267-8

Shi M., H. Gong, L. Cui, Q. Wang, C. Wang, Y. Wang, and G. Kai. 2020. "Targeted metabolic engineering of committed steps improves anti-cancer drug camptothecin production in *Ophiorrhiza pumila* hairy roots." *Industrial Crops and Products* 148: 112277. https://doi.org/10.1016/j.indcrop.2020.112277

Siham A. S., A. A. Ahmed, A. Muhammad, and L. Umme. 2021. "Plant tissue cultural technique to increase production of phytochemicals from medicinal plants: A review." *Plant Archives* 21(1): 1224–1229. https://doi.org/10.51470/PLANTARCHIVES.2021.v21.S1.193

Stefanowicz-Hajduk J., M. Gucwa, B. Moniuszko-Szajwaj, A. Stochmal, A. Kawiak, and J. R. Ochocka. 2021. "Bersaldegenin-1,3,5-orthoacetate induces caspase-independent cell death, DNA damage and cell cycle arrest in human cervical cancer HeLa cells." *Pharmaceutical Biology* 59(1): 54–65. https://doi.org/10.1080/13880209.2020.1866025

Sui Y., L. Jia-Xi, Z. Yue, G. Wen-Hua, D. Jin-Ling, and Y. Xiang-Ling. 2022. "A suspension culture of the hormone autotrophic cell line of *Aralia elata* (Miq.) Seem. for production of oleanolic acid and flavonoids." *Industrial Crops and Products* 176: 114368.

Suryawanshi, S., P. Kshirsagar, P. Kamble, V. Bapat, and J. Jadhav. 2022. "Systematic enhancement of L-DOPA and secondary metabolites from Mucuna imbricata: Implication of precursors and elicitors in callus culture." *South African Journal of Botany* 144: 419–429. https://doi.org/10.1016/j.sajb.2021.09.004

Sykłowska-Baranek K., G. Sygitowicz, A. Maciejak-Jastrzębska, A. Pietrosiuk, and A. Szakiel. 2022. "Application of priming strategy for enhanced paclitaxel biosynthesis in Taxus× Media hairy root cultures." *Cells* 11(13): 2062. https://doi.org/10.3390/cells11132062

Teoh E. S. 2016. "Secondary metabolites of plants." *Medicinal Orchids of Asia* 59–73. https://doi.org/10.1007/978-3-319-24274-3_5

Urquiza-López A., G. Álvarez-Rivera, D. Ballesteros-Vivas, A. Cifuentes, and A. A. Del Villar-Martínez. 2021. "Metabolite profiling of rosemary cell lines with antiproliferative potential against Human HT-29 colon cancer cells." *Plant Foods for Human Nutrition* 76(3): 319–325. https://doi.org/10.1007/s11130-021-00892-w

Vargas-Morales N., N. E. Moreno-Anzúrez, J. Téllez-Román, I. Perea-Arango, S. Valencia-Díaz, A. Leija-Salas, E. R. Díaz-García, P. Nicasio-Torres, M. D. C. Gutiérrez-Villafuerte, J. Tortoriello-García, and J. Arellano-García. 2022. "Spontaneous regeneration of plantlets derived from hairy root cultures of *Lopezia racemosa* and the cytotoxic activity of their organic extracts. *Plants* 11(2): 150. http://doi.org/10.3390/plants11020150

Vinceković M., M. Viskić, S. Jurić, J. Giacometti, D. B. Kovačević, P. Putnik, F. Donsi, F. J. Barba, and A. R. Jambrak. 2017. "Innovative technologies for encapsulation of Mediterranean plants extracts." *Trends in Food Science and Technology* 69: 1–12. https://doi.org/10.1016/j.tifs.2017.08.001

Wang M. R., Z. H. Cui, J. W. Li, X. Y. Hao, L. Zhao, and Q. C. Wang. 2018. "*In vitro* thermotherapy-based methods for plant virus eradication." *Plant Methods* 14: 87. https://doi.org/10.1186/s13007-018-0355-y

Yamamoto S., Y. Murakami, S. Hayashi, and H. Miyasaka. 2022. "Production of paclitaxel in a plant cell culture by in situ extraction with sequential refreshment of water-immiscible 1-butyl-1-methylpyrrolidinium Bis(trifluoromethanesulfonyl)imide." *Solvent Extraction Research and Development Japan* 29(2): 73–78. https://doi.org/10.15261/serdj.29.73

Younessi-Hamzekhanlu M., Z. Dibazarnia, S. Oustan, T. Vinson, R. Katam, and N. Mahna. 2021. "Mild salinity stimulates biochemical activities and metabolites associated with anticancer activities in black horehound (*Ballota nigra* L.). *Agronomy* 11: 2538. https://doi.org/10.3390/agronomy11122538

Zhang B., M. Chen, S. Pu, L. Chen, X. Zhang, J. Zhang, and C. Zhu. 2020. "Identification of secondary metabolites in *Tripterygium wilfordii* hairy roots and culture optimization for enhancing wilforgine and wilforine production." *Industrial Crops and Products* 148: 112276. https://doi.org/10.1016/j.indcrop.2020.112276

Zhang M., A. Wang, M. Qin, X. Qin, S. Yang, S. Su, Y. Sun, and L. Zhang. 2021. "Direct and indirect somatic embryogenesis induction in *Camellia oleifera* Abel." *Frontiers in Plant Science* 12: 644389. https://doi.org/10.3389/fpls.2021.644389

Zhang X., X. Lu, S. Ma, J. Bao, X. Zhang, P. Tian, J. Yang, Y. Lu, and S. Li. 2021. "Research on the release mechanism of glucoraphanin and sulforaphane mediated by methyl jasmonate in broccoli hairy roots." *In Vitro Cellular and Developmental Biology – Plant* 57: 831–841. https://doi.org/10.1007/s11627-021-10225-w

Secondary Metabolites of Mesoamerican Plants *Castilleja tenuiflora* Benth. and *Baccharis conferta* Kunth with Anti-inflammatory Effects

Norma Elizabeth Moreno-Anzúrez,
José Luis Trejo-Espino, Elizabeth Rubio-Rodríguez,
Ana Silvia Gutiérrez-Román,
and Gabriela Trejo-Tapia

CONTENTS

DOI: 10.1201/9781003166535-2

2.1 INTRODUCTION

According to the World Health Organization (WHO), traditional medicine is defined as

> the total amount of knowledge, capacities, and practices based on theories, beliefs, and experiences of different cultures, whether explainable or not, used to maintain health and prevent, diagnose, improve or treat physical and mental diseases
>
> (WHO 2013).

At present, the use of traditional medicine has expanded globally and, in many cases, is the main basis for the provision of health services or their complement. The use of plants has an important role, such that the WHO has reported that about 80% of the world's population uses herbs for their primary health care, and has established three types of medicines, namely: raw plant material, processed plant material, and plant-derived products (WHO 2005).

The term 'Mesoamerica' was originally used by Kirchhoff (1943) to refer to an area where different indigenous groups had developed and shared common cultural characteristics. This region does not have defined boundaries and it ranges from the central part of Mexico to the northwestern part of Costa Rica (Matos-Moctezuma 1994). The Mesoamerican region comprises specific geographical characteristics, including a wide variety of ecosystems and habitats, a broad diversity of climates, and consequently, abundant biodiversity and endemism. This bio-cultural feature is exceptional and a determining factor for the generation of important knowledge about the use of medicinal plants that has been occurring in the region over hundreds of years.

On the other hand, inflammation is a complex biological process, the purpose of which is to maintain cell homeostasis and protect the organism from multiple factors such as physical damage, infections by pathogens, or chemical agents, among others. This process plays a key role in the initiation and development of many chronic degenerative diseases, like Alzheimer's disease, arthritis, cancer, or diabetes. Therefore, there is a wide interest in discovering molecules with anti-inflammatory activities, elucidating their mechanisms of action, and developing strategies to produce them.

In this regard, plants represent an important resource for treating inflammation-related diseases. The pharmacological attributes of plants are related largely to their secondary metabolites.

Compounds with different chemical structures can be complex, and they can come from different metabolic pathways (Chandran et al. 2020). Nowadays, plants continue to represent an important raw material for obtaining secondary metabolites with anti-inflammatory activity. This biological activity has been found in plants of wild origin, as well as those cultivated or generated *in vitro*. Plant tissue culture has been used successfully in the production of secondary metabolites with anti-inflammatory activity by using tools such as micropropagation, suspension cell cultures, metabolic engineering, and bioreactors, thereby increasing the production of the compounds responsible for this activity in high concentrations and under controlled conditions that ensure stable production.

This chapter discusses the potential of the Mesoamerican plants *Castilleja tenuiflora* Benth. ("cancer grass", "hit grass") and *Baccharis conferta* Kunth ("brush", "broom of the mountain") for the treatment of inflammation, and diseases involving this process. It is focused on the progress made in obtaining secondary metabolites, their increased production under controlled conditions (*in vitro*), their chemical identification and elucidation, their biological activity, and studies related to the corresponding biosynthetic pathways.

2.2 INFLAMMATION

Inflammation is an essential biological response that plays an important role in the development of various diseases, and it is triggered by infections, tissue injuries, or genetic changes (Netea et al. 2017; Olajide and Sarker 2020). Inflammatory diseases involve a large number of molecular, immunological, and physiological processes, mainly through the activation of signaling cascades (phosphoinositol kinase and protein kinase C), activation of transcription factors (NF-κB), increased levels of inflammatory enzymes (cyclooxygenase-1, COX-1, and cyclooxygenase-2, COX-2), and release of inflammatory mediators (nitric oxide, NO; prostaglandin E2, PGE2, and cytokines such as interleukin-1 β, IL-1β, and tumor necrosis factor, TNF) (Malik et al. 2017; Netea et al. 2017; Pan et al. 2010;

Zhao et al. 2019). A dysregulated inflammatory response causes inflammatory, autoimmune, and metabolic diseases (Mollaei et al. 2020).

Oral administration of glucocorticoids, non-steroidal anti-inflammatory drugs (NSAIDs, e.g., aspirin, ibuprofen, indomethacin, and naproxen) and specific COX-2 inhibitors (e.g., rofecoxib, celecoxib, lumiracoxib, and parecoxib) are effective for treating inflammation. However, they can have serious side effects on the gastrointestinal system, the heart, and the brain (Seo et al. 2021). For these reasons, research on natural products with anti-inflammatory activity that inhibit this process at different levels can be an effective and possibly safer tool for the treatment of inflammatory diseases, which is why research on natural products with anti-inflammatory activity has been increasing in recent years (Malik et al. 2017).

Some secondary plant metabolites, such as alkaloids, flavonoids, and terpenes, have been shown to have anti-inflammatory properties through different mechanisms in animal models and *in vitro* (Maleki et al. 2019). The anti-inflammatory mechanisms reported are suppression of the release of IL-16, TNF, and NO as well as inhibition of COX-2, inductive NO synthase (iNOS), and NF-κB activity (Ginwala et al. 2019).

2.3 MESOAMERICA AS A SOURCE OF PLANTS WITH ANTI-INFLAMMATORY EFFECTS

Mesoamerica extends from central Mexico to the northwestern part of Costa Rica. Named by the ethnologist and linguist Paul Kirchhoff, this term was used to reflect the geographical similarities and the political, social, economic, and cultural organization of the region (Alonso-Castro et al. 2015). Valuable literary resources contain information on how to use plants for the health care of Mesoamerican inhabitants. Among those currently available are: the Florentine Codex "General History of the Things of New Spain" (Fray Bernardino de Sahagún); the Codex De La Cruz-Badiano (Martín de la Cruz and Juan Badiano) "*Libellus de medicinalibus indorum herbis*", that illustrates 185 plants and mentions 227 names of plants; "*Historia Natural de la Nueva España*" (Francisco Hernández de Toledo), which is a compilation of more than 3,000 plant species, with descriptions of morphology, ecology, and geographical data, and which makes reference to the therapeutic use of some of those plants; the "Alphabetical Catalog of Vulgar [Common] and Scientific Names of Plants that Exist in Mexico" (Martínez 1944); "The Useful Plants of the Mexican Republic" (1928); and "Medicinal Plants of Mexico" (1933). Other important documents include the Mesoamerican

Codices illustrated without texts; however, most of them were destroyed (Bay and Linares 2016).

At present, ethnobotany is a field of research that focuses on documenting and understanding the interaction between humans and plants. It is worth mentioning that Mesoamerica is an extremely rich region of plant sources with bioactive compounds (Casas et al. 2016; de Sousa Araújo et al. 2016). Plants are part of the cultural and traditional knowledge of indigenous communities. In recent years, many ethnobotanical studies have been conducted and have recorded the great cultural and biological diversity of this region (Alonso-Castro et al. 2015). These studies allow a greater understanding of the role of plants among the cultures of the past and are the basis for the selection of pharmacological and phytochemical investigations in the search for new bioactive molecules.

Plants have been used to treat all types of illnesses, ranging from infectious agents such as bacteria or fungi to metabolic and neurological disorders. Therefore many plants have been used traditionally for the treatment of inflammation, and inflammation-related diseases (Bernstein et al. 2018). The concepts and etiology of inflammation, in traditional medicine are complex and do not always correspond with those of modern medicine. There are many conditions related to inflammation, like pain, swelling, suppuration, and production of body fluids, that are treated with different medicinal plants prepared in various ways, such as decoctions, infusions, tinctures, and poultices, among others, and, in most cases, more than one species of plant is used for its treatment.

Most studies on ethnobotany and the pharmacological effects of Mesoamerican plants have been reported recently. Most of these studies are based on the use of *in vivo* or *in vitro* models, although few have included clinical trials. The most used *in vivo* model is ear edema induced by some irritant agent, such as 12-*O*-tetradecanoylphorbol acetate (TPA), croton oil, xylene, or arachidonic acid. This model involves measuring the percentage of inflammation inhibition promoted by plant extracts, fractions, or compounds. The following plants have been reported to inhibit induced ear edema: *Bouvardia ternifolia* Schltdl. (García-Morales et al. 2015), *Buddleja cordata* Kunth (Gutiérrez-Rebolledo et al. 2018), *Bursera copallifera* (Sessé & Moc.) Bullock (Romero-Estrada et al. 2016), and *Baccharis conferta* Kunth (Gutiérrez-Román et al. 2020). Another *in vivo* model used to assess this effect is carrageenan-induced subplantar edema, which has been used for the evaluation of the following plants: *Agave angustifolia* var. *Marginata* (El-Hawary et al. 2020), *B. cordata* (Gutiérrez-Rebolledo et al. 2018), *Cuphea aequipetala* Cav. (Alonso-Castro et al. 2021), *Syngonium podophyllum* Schott (Hossain et al. 2017), *Hamelia patens* Jacq. (Jiménez-Suárez et al. 2016), and *Galphimia glauca* Cav. (Rao et al. 2019).

With *in vitro* models, the stimulation of macrophages with lipopolysaccharides (LPS) derived from bacteria has been used to assess the release of inflammatory mediators. The ethanolic extracts of leaves and stems of *C. aequipetala* decreased the production of NO and hydrogen peroxide (H_2O_2) in macrophages treated with LPS in a dose-dependent response. However, they had low inhibitory activity toward TNF-α, IL-6, and IL-1β, and promoted an increase in IL-10, which is an anti-inflammatory mediator (Alonso-Castro et al. 2021).

2.4 CURRENT APPROACHES TOWARD *IN-VITRO* PRODUCTION OF ANTI-INFLAMMATORY COMPOUNDS

Plant tissue culture is a major component of biotechnology in plants, and it is mainly used for those species that are threatened with extinction, have a low yield or slow growth, and are susceptible to biotic stress (Niazian 2019). It is a promising alternative to producing secondary metabolites of commercial value, mainly for those complex molecules for which chemical synthesis is not economically viable (Roychowdhury et al. 2016). Cell, tissue, and organ cultures also offer the possibility of producing bioactive molecules in a stable and controlled manner, with the chances of increasing their concentrations by adding elicitors or by scaling up the cultures. These biotechnological tools have been implemented to produce bioactive compounds with anti-inflammatory activity obtained from Mesoamerican plants. For example, callus cultures of *G. glauca* produce high levels of galphimine E, which is one of the compounds responsible for the anti-inflammatory effects of this species (Sharma et al. 2018; Jiménez-Arellanes and Pérez-González 2021). Cell suspension cultures of *Sphaeralcea angustifolia* G. Don have been developed to improve the production of anti-inflammatory compounds, such as scopoletin, tomentin, and sphaeralcic acid (Nicasio-Torres et al. 2017). Subsequently, the capacity of *S. angustifolia* cell cultures to produce bioactive compounds was evaluated in a stirred tank bioreactor, and the production of the active metabolites was two times higher in the bioreactor than in the cultures developed in flasks (Pérez-Hernández et al. 2019).

Litsea glaucescens Kunth is considered to be an important species for its use in gastronomy, religious activities, and in traditional Mexican medicine; however, due to these uses and environmental factors, this species is in danger of extinction and is protected by Mexican law (NOM-059-SEMARNAT-2010). In this regard, somatic embryogenesis has been a key strategy to establish *in vitro* regeneration systems for the study and conservation of this species (Dávila-Figueroa et al. 2016). This is of particular importance due to the anti-inflammatory properties that have been demonstrated to be attributable to its flavonoid content (López-Caamal and Reyes-Chilpa 2021).

The cultivation of hairy roots derived from natural genetic transformation mediated by *Agrobacterium rhizogenes* has been widely used to produce secondary metabolites with pharmacological properties because hairy roots grow rapidly in growth regulator-free medium, have high genetic stability, and exhibit high levels of secondary metabolite production (Shi et al. 2021). Transformed roots of S. *angustifolia* were developed to establish root lines with a high-yielding capacity of scopoletin, tomentin, and sphaeralcic acid as the basis for industrial production, which can be stimulated by abiotic factors (nitrates and copper) to produce scopoletin and spheroic acid (Reyes-Pérez et al. 2021). Bazaldúa et al. (2019) reported high levels of podophyllotoxin production in a hairy root line *Hyptis suaveolens* (L.) Poit., induced by *A. rhizogenes* ATCC 15834 + pTDT strain, which produced 100 times more podophyllotoxin than that produced in roots of wild-harvested plants, and 56 times more than the non-transformed roots grown *in vitro.* in the rate of solasodine production in hairy root cultures of *Solanum erianthum* D. Don has been reported to be 1.65 times greater than that produced by the leaves of this species (Sarkar et al. 2020), whereas it was reported that hairy roots of *Lopezia racemosa* Cav. produced terpene-type compounds that had not previously been reported in wild plants or *in vitro.* In addition, the anti-inflammatory activity of hairy root preparations was higher than in wild-harvested plants (Moreno-Anzúrez et al. 2017).

Within the biotechnological strategies used for micropropagation, the use of temporary immersion systems (TIS) has proved to be highly reliable. For instance, the micropropagation efficiency of *A. angustifolia* increased in TIS compared with the conventional tissue culture system (Monja-Mio et al. 2021) and maintained the capacity to produce anti-inflammatory compounds such as phytosterols, saponins, phenolic compounds, and terpenes (Álvarez-Chávez et al. 2021).

Understanding the biosynthetic pathways and the regulation of pharmacologically important metabolites is essential to achieve the highest production of bioactive compounds through approaches such as synthetic biology, tissue engineering, and genetic engineering (Rai et al. 2017). In recent years, the use of these approaches has increased due to the greater number of genomic resources available and the development of new analytical and computational tools. Such is the case of *Mirabilis jalapa* L., which produces antioxidant and anti-inflammatory compounds. Studies of its metabolic and transcriptome profiles have provided important information about the biosynthesis, regulation, and evolution of the pathways of betalains and flavonoids. In *M. jalapa*, betalain-related genes such as cytochrome P450 *(CYP76AD15)* and hydroxycinnamate glucosyltransferase *(MjHCGT)* were identified and functionally characterized. Besides, the expression of genes of the flavonoid/anthocyanin routes was observed; specifically, anthocyanidin synthase (*MjANS*) was highly expressed in betalain-producing organs, but the transcript presented a deletion of 69 important amino acids corresponding to the active site of the enzyme (part of the domain of the oxygenase 20G-Fe (II)), indicating a possible event of elimination of genes that occurred during the evolution of *M. jalapa* and providing new knowledge about evolutionary events leading to betalains production (Polturak et al. 2016; Polturak et al. 2018).

2.5 CASTILLEJA TENUIFLORA BENTH.

Castilleja tenuiflora is described in the book *"Historia de las Plantas de la Nueva España"* by Francisco Hernández (a sixteenth-century naturalist) as a plant of calorific nature, known as 'Atzoyatl' (illacatzic or twisted) (http://www.franciscohernandez.unam.mx/tomos/02_TOMO/tomo002_002/tomo002_002_051.html). It is also known by the following common names: "bella Inés", "Indian shorts", "Castilleja", "sheep's tail", "rabbit", "crane top-knot", "enchiladitas", "ice flower", "cancer grass", "hit grass", "little bananas" or "bloodthirsty" (Martínez 1944). Traditional uses recorded since the sixteenth century include as an antispasmodic, for the treatment of stomach diseases, and as a blood purifier. In the twentieth century, Martínez (1969) described its use to treat anemia, hepatic colic, and gallbladder stones. It was also recommended to treat cancer problems, such as tumors, infertility, cirrhosis, and inflammation, as well as respiratory, and gastrointestinal problems (Béjar et al. 2000; Graham et al. 2000; Alonso-Castro et al. 2011).

C. tenuiflora is a hemiparasitic plant that belongs to the Orobanchaceae family. It is herbaceous with a terrestrial growing in pine and oak forests at altitudes from 740 to 3400 m above sea level. It is a generalist hemiparasite that has many hosts (more than 100). The main hosts that have been reported for *C. tenuiflora* are *Abies religiosa* (Kunth) Schltdl. & Cham.

(Pinaceae), *Bidens triplinervia* Kunth (Asteraceae), *Lupinus montanus* HBK (Fabaceae), *Trisetum spicatum* (L) K. Richt (Poaceae), and *Baccharis conferta* Kunth (Asteraceae). With the latter, the plant establishes an interaction at the root-root level through haustoria (Montes-Hernández et al. 2015). Haustoria are organs that develop in the roots of the parasitic plant and allow its adhesion to and invasion of root cells, and absorption of nutrients and water from the host plant (Yoshida et al. 2016). The haustoria of *C. tenuiflora* is globular, with hairs, and exhibit dermal, fundamental, vascular, and young endophytes. They develop mainly in the vegetative stage and the lateral roots, and exhibit three main tissues, namely dermal, fundamental, and vascular. The composition of haustoria is characterized by the accumulation of starch and tannins, mainly in young haustoria. The number of haustoria associated with a root can range from one to six, with a diameter of 0.5 to 2.5 mm. However, these values depend on the size of the plant, and the main function of haustoria is related to the efficiency of water transport and mechanical support (Montes-Hernández et al. 2015, 2019).

C. tenuiflora synthesizes secondary metabolites that belong to the following chemical groups: flavonoids, lignans, phenylethanoids, and glycosylated iridoids (Figure 2.1a). The flavonoids reported for this plant are flavones (apigenin, luteolin 7-methyl ether, and luteolin 5-methyl ether) and flavonols (quercetin 3-β-D-glucoside and quercetin 3-β-D-rutinoside) (López-Laredo et al., 2012; Rubio-Rodríguez et al. 2019; López-Rodríguez et al. 2019). In the case of lignans, recent studies of *C. tenuiflora* have reported those of the furofuran-type, such as tenuifloroside (glycosylated), sesamin, eudesmin, magnolin, and kobusin (not glycosylated) (Herrera-Ruíz et al. 2015; Rubio-Rodríguez et al. 2019; Arango-de la Pava et al. 2021). The main glycosylated phenylethanoids produced by *C. tenuiflora* are verbascoside and its isomer isoverbascoside. They are structurally characterized by having caffeic acid (a phenylpropanoid residue) and 4,5 hydroxyphenyl ethanol (a phenylethanoid residue) linked to a ß-(D)-glucopyranoside through an ester and a glycosidic bond, respectively. In these molecules, rhamnose is linked in sequence to glucose (Gómez-Aguirre et al. 2012).

In the iridoid group, the main compound reported is aucubin. Other iridoids found in *C. tenuiflora* are geniposidic, mussaenoside, shanzhiside acids, bartsioside, and carioptoside (Jiménez et al. 1995; Carrillo-Ocampo et al. 2013; Herrera-Ruíz et al. 2015; Cortes-Morales et al. 2018; Rubio-Rodríguez et al. 2021).

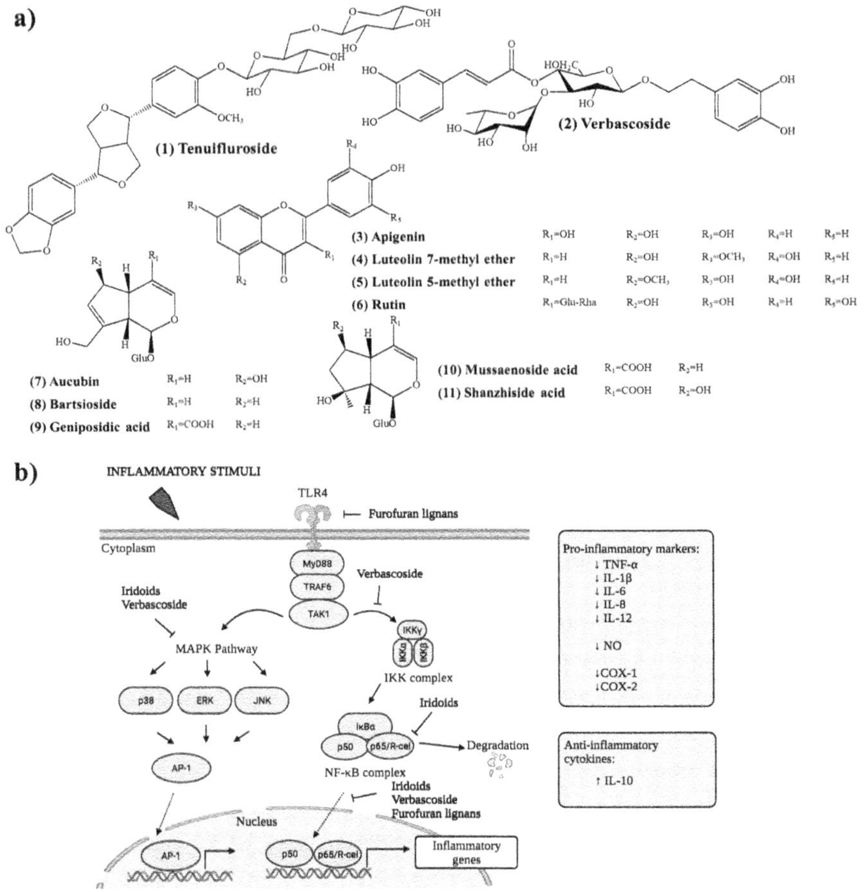

FIGURE 2.1 *Castilleja tenuiflora* (a) Chemical structures of metabolites synthesized by *C. tenuiflora.* (b) Inflammatory signaling pathways inhibited by the pharmacological active compounds of *C. tenuiflora.* Extracellular stimuli activate TLR receptors triggering intracellular signaling cascades that lead to nuclear translocation of AP-1 and NF-κB via the MAPK pathway or NF-κB pathways. AP-1 and the heterodimer p50-p65 regulate the expression of inflammatory mediators like TNF-α, IL-1β or iNOS. Figure 2.1b was created with Biorender.com.

C. tenuiflora extracts have pharmacological activities, such as cytotoxic (Moreno-Escobar et al. 2011), antioxidant (Moreno-Escobar et al. 2011; López-Laredo et al. 2012), anti-inflammatory (Carrillo-Ocampo et al. 2013; Sanchez et al. 2013; Arango-de la Pava et al. 2021), anti-ulcerogenic (Sanchez et al. 2013), antidepressant (Herrera-Ruiz et al. 2015; López-Rodríguez et al. 2019), and sedative effects (Herrera-Ruiz et al.

2015) (Table 2.1). The antioxidant activity of root and shoot cultures has been recorded *in vitro*, with EC_{50} doses of 6 and 1.3 µg methanolic extract/mg DPPH, respectively. This effect was associated with the concentration of phenolic compounds and flavonoids present in the plant (Trejo-Tapia et al. 2012).

The anti-inflammatory activity of *C. tenuiflora* extracts has been reported, mainly from ethanolic, methanolic, ethyl acetate, and aqueous extracts. Carrillo-Ocampo et al. (2013) reported that the fraction with the highest abundance of iridoids (aucubin, barsioside, geniposidic acid, and mussaenoside acid) and verbascoside had higher anti-inflammatory activity in a model of mouse ear edema induced by TPA. Iridoids (aucubin, geniposidic acid, and mussaenoside acid) exhibited a similar activity to that of indomethacin (the reference control): the percentage inhibition was 71.54, 91.01, and 65.34% of the indomethacin control, respectively (at a dose of 0.1 mg/ear). On the other hand, loganic acid, 8-*epi*-loganin, and geniposide exhibit anti-inflammatory activity in similar percentage inhibition (79.82, 81.14, and 86.61%, respectively). Sanchez et al. (2013) reported the anti-inflammatory activity of ethyl acetate extracts and aqueous extracts from wild-harvested *C. tenuiflora* plants and from material obtained *in vitro*, also using the model of ear inflammation induced by TPA. These authors demonstrated that the extracts promoted moderate anti-inflammatory activity, attributed mainly to verbascoside and aucubin, with 38.1 and 49.1% inhibition (1.6 mg/ear), below dexamethasone with 57.7% (1.6 mg/ear). It was possible to obtain a fraction with a higher concentration of bartsioside from the methanolic extract. It had a strong anti-inflammatory activity, with 70% inhibition at a dose of 1.5 mg/ear, similar to dexamethasone.

Recently, it was reported that binary mixtures of the compounds produced by *C. tenuiflora* promoted pharmacodynamic interactions in the inhibition of the NF-κB/AP-1 pathway. RAW-Blue™ cells were used to assess eight compounds isolated from *C. tenuiflora*, i.e., five furofuran-like lignans (tenuifloroside, magnolin, eudesmin, sesamin, and kobusin), one phenylethanoid (verbascoside), and two iridoids (geniposide and aucubin). Tenuifloroside, magnolin, eudesmin, sesamin, and kobusin exhibited a percentage inhibition of the NF-κB/AP-1 pathway of 52.91, 60.97, 56.82, 52.63, and 45.45%, respectively. Verbascoside had a percentage inhibition of 55.55%, whereas the iridoids (aucubin and geniposide) had an inhibition of 40.00 and 57.47%, respectively. Verbascoside was chosen as the leading compound to make the binary mixtures, i.e., verbascoside-lignans

TABLE 2.1 Ethnobotanic Use, Phytochemistry, Anti-inflammatory Activity and *in vitro* Culture of *Castilleja tenuiflora* and *Baccharis conferta*

Botanical family ***Species*** **Ethnobotanic use**	**Anti-inflammatory activity (% inhibition/dose)**	***In vitro* culture**	**Reference**
Orobanchaceae ***Castilleja tenuiflora* Benth** Leaf and flower decoction used to treat symptoms of cancer, sterility, gastrointestinal disorders, cirrhosis, inflammation, and respiratory diseases	Model: TPA-induced mouse ear edema. Geniposidic acid (91.01% inhibition), geniposide (86.61%), 8-*epi*-loganin (81.14%), loganic acid (79.82%), aucubin (71.54%), musssaenoside (69.02%) at 0.1 mg/ear. Pharmacodynamic interactions in inhibition of the NF-κB/AP-1 pathway using RAW-Blue™ cells. Binary mixtures: Verbascoside-aucubin, verbascoside-kobusin (synergism) and verbascoside-tenuifloroside (subadditivity)	Shoot induction in MS with IBA (0.5 µM), adventitious root cultures in B5 with IAA (10 µM) or NAA (10 µM), shoot elicitation with CWOs (13 µg/mL) from *Fusarium oxysporum* f. sp. *lycopersici* race 3, synthetic seeds, haustorium induction with catechin, vanillin or H_2O_2 (25 µM), culture in RITA® + nitrogen deficiency as stimulus	Carrillo-Ocampo et al. 2013; Arango-de la Pava et al. 2021; Martínez-Bonfil et al. 2011; Gómez-Aguirre et al. 2012; Cardenas-Sandoval et al. 2015; León-Romero et al. 2019; Salcedo-Morales et al. 2017; Valdez-Tapia et al. 2014; Cortes-Morales et al. 2018
Asteraceae ***Baccharis conferta* Kunth** Decoction or infusion prepared from aerial parts to treat joint pain, seizures, cramps, toothache, colds, and digestive disorders	Model: TPA-induced mouse ear edema. Schensianol A (78.18% inhibition), bacchofertin (83.50%), kingidiol (94.13%), bacchofertone (76.78%), cirsimaritin (98.14%), hispidulin (74.41%) at 1 mg/ear	Shoot multiplication from nodal segments in MS with IBA (5 µM) + KIN (5 µM) or IBA (0.5 µM) + BAP (1.10 µM). Friable calli using TDZ (5 µM)	Gutiérrez-Román et al. 2020; Leyva-Peralta et al. 2019

BAP: benzylaminopurine; B5: Gamborg´s B5 medium; KIN: kinetin; IBA: indole butyric acid; MS: Murashige and Skoog medium; NAA: naphthalene acetic acid; TDZ: thidiazuron; CWOs: cell wall oligosaccharides and verbascoside-iridoids. The verbascoside-lignans mixtures showed a wide range of NF-κB/AP-1 inhibition values; verbascoside-kobusin showed the highest inhibition of NF-κB/AP-1, (63.06%), followed by verbascoside-eudesmin (51.01%), verbascoside-sesamin (47.40%), and verbascoside-magnolin (39.33%), whereas a verbascoside-tenuifloroside mixture showed the lowest value (37.99%). Aucubin exhibited a synergistic effect when mixed with verbascoside, indicating that the inhibition of NF-κB/AP-1 increased when the concentration of verbascoside increased relative to aucubin. On the other hand, a sub-additive effect was observed when the iridoid concentration was higher than that of verbascoside, indicating that the higher the iridoid concentration, the lower the NF-κB/AP-1 inhibition (Arango de la Pava et al. 2021).

Iridoids occupy an important position in the field of natural product chemistry and pharmacology. For example, iridoids such as catalpol, geniposide, geniposidic acid, and loganin have been reported to inhibit COX-1 and COX-2, the enzymes involved in part of the anti-inflammatory process (Park et al. 2007; Shi et al. 2013). Loganin and geniposide exhibited inhibitory activity for COX-1, with half-maximal inhibitory concentration (IC_{50}) values of 3.55 and 5.37 mM, respectively. Catapol, geniposide, loganin, and aucubin suppressed TNF-α levels, with IC_{50} values of 11.2, 33.3, 58.2, and 154.6 mM, respectively (Park et al. 2007). Geniposide also inhibited nitric oxide production and regulated the central role of TNF-α through the control of pro-inflammatory cytokines. In addition, it has been reported that the inhibitory activity of geniposide resulted from the inhibition of the phosphorylation of p38, Erk1/2, and JNK, through the AP-1 and NF-κB pathways, by the degradation of IκB by macrophages (Shi et al. 2013).

On the other hand, aucubin exhibits various biological activities and promises to be a safe and potent drug for treating various diseases (Zeng et al. 2020). Regarding anti-inflammatory biological activity, it was reported that this compound participated in the metabolism of arachidonic acid. It is an important compound since, in its aglycone form, it inhibits COX-2 (Wang et al. 2015). It was also shown that aucubin is involved in the regulation of inflammatory mediators, reducing the production of IL-1β and IL-6 (Kang et al. 2018), blocking the production of TNF-α, and inhibiting the activation of NF-κB (Qiu et al. 2018). Similarly, it has been reported that the inhibitory activity of aucubin via NF-κB/AP-1 was due to the suppression of the degradation of the κB (I κB) inhibitor and the translocation of the p65 subunit (Jain et al. 2016) (Figure 2.1b).

The potent anti-inflammatory activity of verbascoside has been mainly attributed to the inhibition of pro-inflammatory molecules at the transcriptional level, involving the COX-2 enzyme, through the inhibition of histamine and arachidonic acid release, the enzyme calcineurin, an important regulator of T-cell-mediated inflammation, and the activator protein 1 (AP-1), an important modulator of inflammatory processes in chronic inflammatory diseases (Yoou et al. 2015). It negatively regulates Ca^{2+}-dependent MAPK signaling (Gao et al. 2017), reduces macrophages that are stimulated by LPS and IFN-γ, and completely inhibits the production of NO, TNF-α, and IL-12 (Han et al. 2018). The process by which it carries out its activity as a potent anti-inflammatory agent is due to its structure as an aromatic hydrocarbon. Verbascoside binds to the transcription factor aryl hydrocarbon receptor (AhR), thus contributing to the transcription of numerous detoxification genes that code for phase I and II drug/toxin-metabolizing enzymes, in particular the subfamily of cytochrome P450 CYP1, Nrf2, glutathione S-transferase (GST), and antioxidant enzymes. As a result, verbascoside inhibits pro-inflammatory cytokines and subsequent growth factors (Tian et al. 2021). On the other hand, it has anti-inflammatory activity on the NF-κB/AP-1 pathway, inhibiting the phosphorylation of TAK-1 through the activation of the protein tyrosine phosphatases (Pesce et al. 2015).

Lignans can decrease the inflammatory factors TNF-α, IL-1β, and interleukin-6 (IL-6), regulated by NF-κB (Zhou et al. 2020). Some lignans negatively regulate the phosphorylation of the IκB kinase IKK complex and increase the phosphorylation of p65 and the nuclear concentration of p65 (Xian et al. 2019). On the other hand, lignans exert a regulatory effect on MAPK in the NF-κB signaling pathway. It was reported that they decreased the phosphorylation of ERK, p38, and JNK, directly inhibiting NF-κB (Yang et al. 2019). They participated in the intercommunication between the Keap1/Nrf2 and NF-κB signaling pathways involved in antioxidant and anti-inflammatory properties (Wardyn et al. 2015). At the transcriptional level, lignans regulate multiple microRNAs, mainly miR-17-5p, miR-145, and miR-127, which are implicated in antioxidant and anti-inflammatory effects (Zhou et al. 2020). The anti-inflammatory activity of furofuran lignans results from the high number of methoxy groups present. Methoxy groups have been reported to be important agents in the anti-inflammatory activity of NF-κB/AP-1, potentially inhibiting inflammation *in vitro* and suppressing LPS-induced phosphorylation of IKK and IκBα degradation (Yang et al. 2017). It has been reported that kobusin inhibited nuclear translocation of the p65 subunit of NF-κB and IκBα phosphorylation (Kim et al. 2009). Sesamin inhibited the expression of an LPS receptor, TLR-4, which initiates the signaling cascade for the activation of NF-κB and AP-1 (Udomruk et al. 2018). The anti-inflammatory activity of NF-κB/AP-1 achieved by furofuran lignans may be due to the inhibitory effect on TLR-4 (Arango-de la Pava et al. 2021).

2.6 *Baccharis conferta* KUNTH

Baccharis conferta is described in the "*Historia de las Plantas de la Nueva España*" by Francisco Hernández as a "hot" plant, known as 'Quauhizquiztli' or 'caratacua' (http://www.ibiologia.unam.mx/plantasnuevaespana/index.html). It is also known by the common names of "broom", "azoyate", "brush" (Morelos), "broad broom", "broom of the mountain" (Veracruz), and "grass of the carbonero" (Mexico Valley), among others. *Baccharis conferta* belongs to the Asteraceae family. It is distributed in the neovolcanic axis at an altitude ranging from 1900 to 3600 m above sea level (Argueta-Villamar et al. 1994; Monroy-Ortíz and Castillo-España 2007). It has been recorded in the ethnobotanical literature that the form of use of this species is by decoctions or infusions of the aerial parts (Heinrich et al. 1998; Mata 2009; Monroy-Ortíz and Castillo-España 2007) to treat joint pain, seizures, cramps, toothaches, colds, and digestive disorders.

The first phytochemical study on *B. conferta* was conducted by Bohlmann and Zdero (1976). The main groups of reported compounds were phenolics and terpenes. In general, flavonoids are the major compounds that have been reported in these species. Through different studies, more than 30 secondary metabolites have been isolated and identified, including sixteen flavonoids, three phenolic acids, two coumarins, three diterpenes, four triterpenes, and two essential oils (Bohlmann and Zdero 1976; Cortes-Morales et al. 2019; Guerrero and Romo del Vivar 1973; Gutiérrez-Román et al. 2020; Weimann et al. 2002) (Figure 2.2a).

The flavonoid profile of *B. conferta* includes flavones, flavanones, and flavonols. Most flavonoids have polyhydroxy substitutions, with a wide variety of *O*-methylation patterns. The presence of the OH group at position 5 of ring A is a common characteristic in most flavonoids described for this species, and mainly *O*-methylations are in the 6, 7, and 4 positions. The flavonoids that present one methoxyl group are apigenin, acacetin, kaempferide, isokaempferide, 6-methoxy kaempferol, and hispidulin. Those with two methoxyl groups are pectolinaringenin, cirsimaritin,

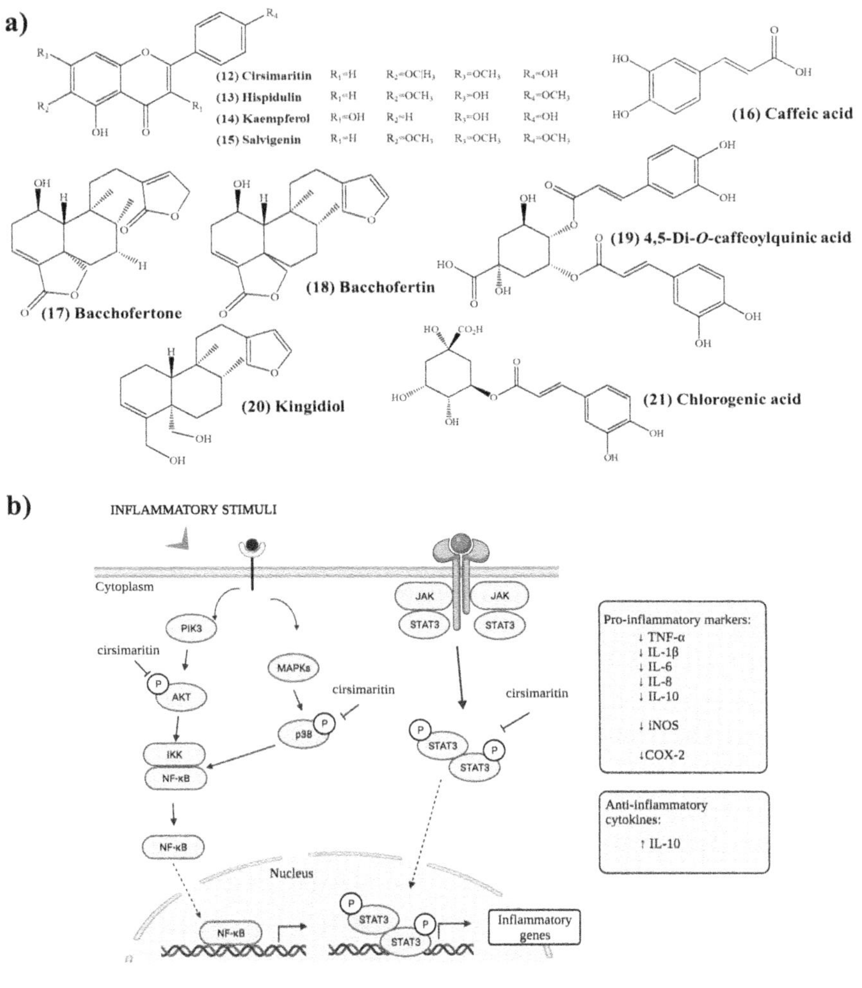

FIGURE 2.2 *Baccharis conferta* (a) Chemical structures of metabolites synthesized by *B. conferta.* (b) Inflammatory signaling pathways inhibited by the pharmacological active compounds of *B. conferta*. Inflammatory signal is transduced by a receptor to activate Janus kinase (JAK) which then activates Signal transducer and activator transcription (STAT). STATs are dephosphorylated in the nucleus leading to activation of inflammatory cytokines. Figure 2.2b was created with Biorender.com.

4'-6-dimethoxy naringenin, dihydro kaempferide, and 4'-7-dimethoxy naringenin, whereas those with three methoxyl groups are salvigenin, eupatilin, 4', 6, 7-trimethoxy kaempferol. Finally, kaempferol and vicenin are two flavonoids that do not exhibit any *O*-methylation (Cortes-Morales et al. 2019; Gutiérrez-Román et al. 2020; Weimann et al. 2002).

Recent studies have reported phenolic acids to be found in *B. conferta*. Caffeoylquinic acids, such as caffeic acid, chlorogenic acid, and 4,5 di-*O*-caffeoylquinic acid are synthesized by wild plants and *in vitro* cultures (Cortes-Morales et al. 2019; Leyva-Peralta et al. 2019).

Triterpenes and diterpenes have also been found to be characteristic of this species, mainly the diterpenes of the neo-clerodane type, such as bacchofertin (Bohlmann and Zdero 1976), kingidiol, and a new diterpene called bacchofertone (Gutiérrez-Román et al. 2020). In addition, triterpenes such as ursolic acid methyl ester, oleanolic acid, and erythrodiol have been isolated from the aerial parts, and baccharis oxide from the roots (Gutiérrez-Román et al. 2020; Weimann et al. 2002). Two essential oils known as caryophyllene and schensianol A have also been reported (Gutiérrez-Román et al. 2020; Weimann et al. 2002) (Figure 2.2a).

Pharmacological studies on the extracts, fractions, and pure compounds of *B. conferta* have identified a relatively wide range of biological activities. It has been reported that this plant has antispasmodic (Weimann et al. 2002) and anti-inflammatory (Gutiérrez-Román et al. 2020) activities, as well as ovicidal effects against the gastrointestinal nematode *Haemonchus contortus* (Cortes-Morales et al. 2019) (Table 2.1).

The pharmacological assessment of the dichloromethane extract and fractions of *B. conferta* indicated anti-inflammatory activity in the model of ear edema induced by TPA in mice, at a dose of 1 mg/ear. The anti-inflammatory activity of six isolated compounds was also assessed, observing a reduction of more than 70% in edema, with their anti-inflammatory activity being dose-dependent. The compounds with percentage edema inhibition greater than 90% were cirsimaritin and kingidiol (Gutiérrez-Román et al. 2020). It has been reported that the flavonoid cirsimaritin inhibits the synthesis of pro-inflammatory cytokines, such as interleukin-6, TNF-α, and NO, by regulating the inhibition of c-Fos and Stat3 phosphorylation in RAW264.7 cells stimulated by LPS (Shin et al. 2017) (Figure 2.2b).

2.7 BIOTECHNOLOGY OF *C. TENUIFLORA* AND *B. CONFERTA*

Due to the potential value of *C. tenuiflora* and *B. conferta* as natural sources of compounds of pharmacological importance, biotechnology systems have been developed to study the production of their secondary metabolites.

For *C. tenuiflora*, the *in vitro* cultures reported regarding the production of these compounds were from roots, shoots, and seedlings (Table 2.1). Systems have also been developed for the conservation of *C. tenuiflora* through the production of synthetic seeds and for the study of its hemiparasite interaction (Martínez-Bonfil et al. 2011; Valdez-Tapia et al. 2014; Salcedo-Morales et al. 2017; León-Romero et al. 2019; Trejo-Tapia et al. 2019) (Figure 2.3a). These systems have mainly been used for enhancing the biosynthesis of chemical compounds of pharmacological interest. In this context, biotic stress has been induced in cultures of *C. tenuiflora* shoots in a liquid medium, by elicitation by exposure to oligosaccharides from the cell wall of *Fusarium oxysporum* at two concentrations, 5.23 and 13.08 μg mL^{-1}. It was observed that the higher concentration increased the concentration of total phenolic compounds, the production of glycosylated phenylethanoids (verbascoside and its isomer), and the activity of one of the enzymes associated with its biosynthesis (Cardenas-Sandoval et al. 2015).

Nitrogen deficiency has been used as a strategy to induce abiotic stress and enhance the production of secondary metabolites in shoots of *C. tenuiflora* using a RITA® reactor system. It was observed that the production of total phenolic compounds, phenylethanoids, and aucubin all increased, and the formation of anthocyanins was also promoted. In addition, there was an increase in the activity of the rate-limiting enzyme of the phenylpropane metabolism pathway, phenylalanine ammonium lyase (PAL) (Medina-Pérez et al. 2015; Cortes-Morales et al. 2018). However, this increase in the production of secondary metabolites was accompanied by a significant decrease in biomass production, in the total concentration of chlorophyll, and multiplication rate and root formation capacity. To avoid this damage, spermine (polyamine) was added to the culture medium, which not only counteracted the observed damage but also generated bartsioside, 8-*epi*-loganin, tenuifloroside, luteolin 5-methyl ether, and quercetin, which had not been detected previously under *in vitro* conditions (Rubio-Rodríguez et al. 2019).

In recent studies, progress has been made in understanding the molecular basis of secondary metabolite biosynthesis in *C. tenuiflora* seedlings. Partial gene sequences that code for four enzymes (DXS, G10H, PAL, and CHS) involved in the biosynthesis of iridoids and phenylpropanoids were isolated and cloned. Subsequently, the *C. tenuiflora* seedlings were elicited by application of three signaling molecules, namely salicylic acid, H_2O_2, or methyl jasmonate (MeJA). The changes in the concentrations of major

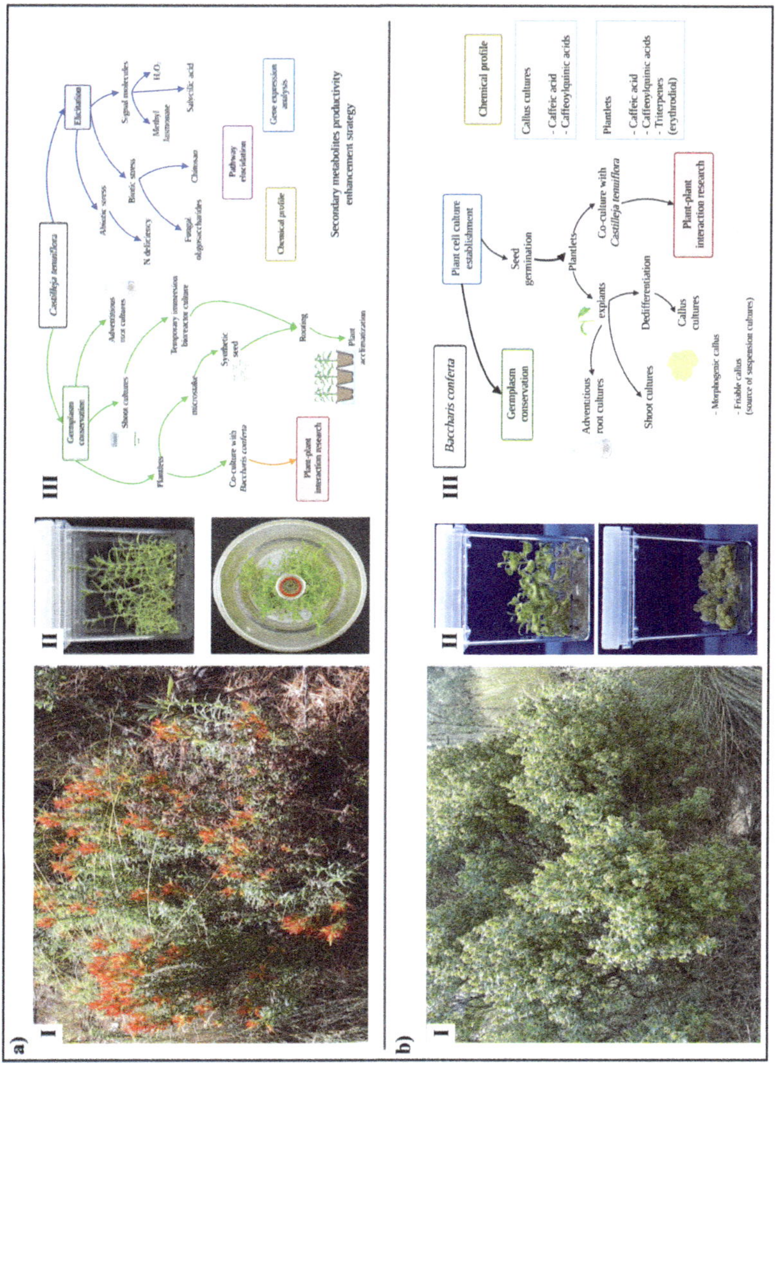

FIGURE 2.3 Plants with anti-inflammatory activity. (a) *Castilleja tenuiflora* I. *C. tenuiflora* in the wild; II. *C. tenuiflora in vitro* seedlings (upper) and plants regenerated in temporary immersion bioreactors (lower); III. Biotechnological strategies of conservation of *C. tenuiflora* germplasm and secondary metabolites productivity enhancement (b) *Baccharis conferta* I. *B. conferta* in the wild; II. *B. conferta in vitro* seedlings (upper) and *in vitro* callus (lower); III. Conservation of *B. conferta* germplasm and production of secondary metabolites with anti-inflammatory activity through biotechnological strategies. Figures 2.3a.III and 2.3b.III were created with Biorender.com.

metabolites (aucubin and verbascoside) and the transcriptional levels of the partial gene sequences were assessed. Salicylic acid regulated the production of the major metabolites (increasing aucubin and decreasing verbascoside) by regulating the transcriptional rate of the genes assessed. The *Cte-PAL1* and *Cte-CHS1* genes were overexpressed under H_2O_2 stimulus, whereas expression of the *Cte-DXs1* and *Cte-G10H* genes was downregulated. These changes promoted a greater accumulation of aucubin than verbascoside. MeJA increased the biosynthesis of the major metabolites and the overexpression of the assessed genes. MeJA is a key elicitor used to stimulate the biosynthesis of metabolites of *C. tenuiflora*, whereas salicylic acid and H_2O_2 can be useful elicitors to assess their regulation (Rubio-Rodríguez et al. 2021).

Given that *C. tenuiflora* is a hemiparasitic plant, *in vitro* cultures have been developed to study the process of haustorium formation. Using an inclined tube system *in vitro*, a series of chemical compounds called haustorium-inducing factors (HIFs) were applied, namely catechin, vanillin, and H_2O_2 (at concentrations of 10, 25, and 50 μM, respectively). Vanillin was the best HIF, inducing three haustoria per root. Phytochemical analysis indicated that the concentration of total phenolics in the aerial part was similar in the vanillin treatment and the control, whereas, in the roots, the concentration was higher in the control group. Loganic acid, verbascoside, and apigenin accumulated to higher concentrations in the root than in the aerial parts and, finally, the histochemical analysis indicated that the formation of haustoria was associated with the accumulation of H_2O_2 and flavonoids. It was concluded that vanillin was an excellent haustorium inducer in *C. tenuiflora*, without the need for the presence of a host plant (Salcedo-Morales et al. 2014).

Regarding *B. conferta*, the procedures for *in vitro* dedifferentiation (callus formation), shoot multiplication, and plantlet regeneration have been established (Leyva-Peralta et al. 2019) (Figure 2.3b). It was observed that explant type and cytokinin/auxin ratio influenced the morphogenic response; this is important to define the culture conditions for the biotechnological system of interest. *In vitro B. conferta* cultures have been characterized for their production of secondary metabolites. Micropropagated seedlings and organogenic calli synthesize caffeic acid, chlorogenic acid, and 4,5 di-*O*-caffeoylquinic acid. The specific yield of caffeoylquinic acids is markedly higher in callus cultures (112.17 mg/g dry matter) compared to *in vitro* seedlings (55.45 mg/g dry matter). Small amounts of the triterpenes erythrodiol and ursolic acid have been observed in seedlings grown *in vitro* (Leyva-Peralta et al. 2019). The chemical profile of *in vitro* cultures of *B. conferta* needs further elucidation.

Furthermore, *in vitro* co-culture system has been used to study the host-parasite interaction between *B. conferta* and *C. tenuiflora*. Under co-culture, the chemical profile of the two species is different compared to axenic cultures. In the root of *C. tenuiflora*, the production of phenylethanoids increased, whereas in *B. conferta*, the concentration of chlorogenic and caffeic acid increased in the aerial parts and in the roots. The molecular mechanisms underlying plant-plant interaction of *B. conferta* and *C. tenuiflora* are under investigation.

2.8 CONCLUSIONS

Over the years, the bio-cultural characteristics of the Mesoamerican region have led to the generation of knowledge about the use of a great variety of native plants. Plants synthesizing secondary metabolites with relevant biological activities have proven to be an invaluable resource for the treatment of chronic degenerative diseases directly related to inflammatory processes. The scientific studies compiled in this chapter indicate that the secondary metabolites synthesized by *C. tenuiflora* and *B. conferta* have outstanding anti-inflammatory activity. Depending on the extract, mixture, or compounds used, those metabolites had effects on different molecular targets, decreasing or inhibiting inflammatory responses.

Plant biotechnology is a method applied through the *in vitro* cultivation of plants to develop systems that can promote the production of secondary metabolites with anti-inflammatory activity. The advances described in this chapter about Mesoamerican plants, specifically *C. tenuiflora* and *B. conferta*, allowed the identification, quantification, and elucidation of chemical structures of secondary metabolites produced by these plants. This progress also enabled the establishment of strategies to enhance *in vitro* production of these plants and has made it possible to measure the changes at the gene transcription level in response to different elicitors, allowing more information to be obtained about the biosynthetic pathways of secondary metabolites with the anti-inflammatory activity of *C. tenuiflora*. *B. conferta in vitro* cultures are a potential source of relevant antioxidants, such as caffeoylquinic acids, and a promising tool to establish secondary-metabolite-enhancing strategies.

Finally, it is essential to take into consideration that, to achieve greater success in the production of plant secondary metabolites *in vitro*, it is necessary to deepen our knowledge and understanding of the relevant biosynthesis pathways and their regulation.

REFERENCES

Alonso-Castro A. J., F. Domínguez, J. R. Zapata-Morales, and C. Carranza-Álvarez. 2015. "Plants used in the traditional medicine of Mesoamerica (Mexico and Central America) and the Caribbean for the treatment of obesity." *Journal of Ethnopharmacology* 175: 335–345. http://doi.org/10.1016/j.jep.2015.09.029

Alonso-Castro A. J., M. L. Villarreal, L. A. Salazar-Olivo, M. Gomez-Sanchez, F. Dominguez, and A. Garcia-Carranca. 2011. "Mexican medicinal plants used for cancer treatment: Pharmacological, phytochemical and ethnobotanical studies." *Journal of Ethnopharmacology* 133(3): 945–972. https://doi.org/10.1016/j.jep.2010.11.055

Alonso-Castro A. J., V. Arana-Argáez, E. Yáñez-Barrientos, M. A. Ramírez-Camacho, K. Wrobel, J. C. Torres-Romero, C. León-Callejas, and K. Wrobel. 2021. "Antinociceptive and antiinflammatory effects of *Cuphea aequipetala* Cav (Lythraceae)." *Inflammopharmacology.* http://doi.org/10.1007/s10787-020-00709-3

Álvarez-Chávez J., M. Villamiel, L. Santos-Zea, and A. K. Ramírez-Jiménez. 2021. "Agave by-products: An overview of their nutraceutical value, current applications, and processing methods." *Polysaccharides* 2(3): 720–743. https://doi.org/10.3390/polysaccharides2030044

Arango-De la Pava L. D., A. Zamilpa, J. L. Trejo-Espino, B. E. Domínguez-Mendoza, E. Jiménez-Ferrer, L. Pérez-Martínez, and G. Trejo-Tapia. 2021. "Synergism and subadditivity of verbascoside-lignans and iridoids binary mixtures isolated from *Castilleja tenuiflora* Benth. on NF-κB/AP-1 inhibition activity." *Molecules* 26(3): 547. http://doi.org/10.3390/molecules26030547

Argueta-Villamar A., L. M. Cano-Asseleih, M. E. Rodarte, and M. C. Gallardo-Vázquez. 1994. *Atlas de las Plantas de la Medicina Tradicional Mexicana: Baccharis conferta 602–3*. Mexico: Instituto Nacional Indigenista Press.

Bay R., and E. Linares. 2016. "Chapter 3: Ethnobotany and ethnohistorical sources of Mesoamerica." In *Ethnobotany of Mexico: Interactions of People and Plants in Mesoamerica*, edited by R. Lira, A. Casas, and J. Blancas, 41–61. New York: Springer Press.

Bazaldúa C., A. Cardoso-Taketa, G. Trejo-Tapia, B. Camacho-Diaz, J. Arellano, E. Ventura-Zapata, and M. L. Villareal. 2019. "Improving the production of podophyllotoxin in hairy roots of *Hyptis suaveolens* induced from regenerated plantlets." *PLoS One* 14(9): e0222464. http://doi.org/10.1371/journal.pone.0222464

Béjar E., R. Reyes-Chilpa, and M. Jiménez-Estrada. 2000. "Bioactive compounds from selected plants used in the XVI century mexican traditional medicine." *Studies in Natural Products Chemistry* 24: 799–844. http://doi.org/10.1016/S1572-5995(00)80055-1

Bernstein N., A. Muhammad, D. Muhammad, H. Koltai, M. Fridlender, and J. Gorelick. 2018. "Antiinflammatory potential of medicinal plants: A source for therapeutic secondary metabolites." *Advances in Agronomy* 131–183. https://doi.org/10.1016/bs.agron.2018.02.003

Bohlmann F., and C. Zdero. 1976. "Weitere inhaltsstoffe aus *Baccharis conferta* H. B. K." *Chemische Berrichte* 109(4): 1450–1452. http://doi.org/10.1002/cber.19761090430

Cardenas-Sandoval B. A., L. Bravo-Luna, K. Bermúdez-Torres, J. L. Trejo-Espino, A. Zamilpa, and G. Trejo-Tapia. 2015. "Enhancement of phenylethanoid glycosides biosynthesis in *Castilleja tenuiflora* Benth. shoot cultures with cell wall oligosaccharides from *Fusarium oxysporum* f. sp. *lycopersici* race 3." *Revista Mexicana de Ingeniería Química* 14(3): 631–639. http://www.scielo.org.mx/scielo.php?script=sci_abstract&pid=S1665-27382015000300007&lng=es&nrm=iso

Carrillo-Ocampo D., S. Bazaldúa-Gómez, J. R. Bonilla-Barbosa, R. Aburto-Amar, and V. Rodríguez-López. 2013. "Anti-inflammatory activity of iridoids and verbascoside isolated from *Castilleja tenuiflora*." *Molecules* 18(10): 12109–12118. http://doi.org/10.3390/molecules181012109

Casas A., J. J. Blancas, and R. Lira. 2016. "Chapter 1: Mexican ethnobotany: Interactions of people and plants." In *Mesoamerica in Ethnobotany of Mexico: Interactions of People and Plants in Mesoamerica*, edited by R. Lira, A. Casas, and J. Blancas, 1–13. New York: Springer Press.

Chandran H., M. Mukesh, T. Barupal, and S. Kanika. 2020. "Plant tissue culture as a perpetual source for production of industrially important bioactive compounds." *Biotechnology Reports* 26: E00450. http://doi.org/10.1016/j.btre.2020.e00450

Cortes-Morales J. A., A. Olmedo-Juárez, G. Trejo-Tapia, M. González-Cortazar, B. E. Domínguez-Mendoza, P. Mendoza-de Gives, and A. Zamilpa. 2019. "*In vitro* ovicidal activity of *Baccharis conferta* Kunth against *Haemonchus contortus*." *Experimental Parasitology* 197: 20–28. http://doi.org/10.1016/j.exppara.2019.01.003

Cortes-Morales J. A., A. R. López-Laredo, A. Zamilpa, K. Bermúdez-Torres, J. L. Trejo-Espino, and G. Trejo-Tapia. 2018. "Morphogenesis and secondary metabolites production in the medicinal plant *Castilleja tenuiflora* Benth. under nitrogen deficiency and starvation stress in a temporary immersion system." *Revista Mexicana de Ingeniería Química* 17(1): 229–242. http://doi.org/10.24275/uam/izt/dcbi/revmexingquim/2018v17n1/Cortes

Dávila-Figueroa C. A., J. F. Morales-Domínguez, M. Rosa-Carrillo, and E. Pérez-Molphe-Balch. 2016. "*In vitro* regeneration of mexican bay (Litsea glaucescens Kunth). Through somatic embryogenesis." *Revista Fitotecnia Mexicana* 39(2): 123–131.

De Sousa-Araújo T. A., J. G. de Melo, W. S. Ferreira-Júnior, and U. P. Albuquerque. 2016. "Medicinal plants." In *Introduction to Ethnobiology*, edited by U. Albuquerque and R. Nóbrega Alves. Cham: Springer. http://doi.org/10.1007/978-3-319-28155-1_22

El-Hawary S. S., H. A. El-Kammar, M. A. Farag, D. O. Saleh, and R. S. El Dine. 2020. "Metabolomic profiling of five Agave leaf taxa via UHPLC/PDA/ESI-MS in relation to their anti-inflammatory, immunomodulatory and ulceroprotective activities." *Steroids* 160: 108648. https://doi.org/10.1016/j.steroids.2020.108648

Gao H., Y. Cui, N. Kang, X. Liu, Y. Liu, Y. Zou, Z. Zhang, X. Li, S. Yang, J. Li, C. Wang, Q. Xu, and X. Chen. 2017. "Isoacteoside, a dihydroxyphenylethyl glycoside, exhibits anti-inflammatory effects through blocking toll-like receptor 4 dimerization." *British Journal of Pharmacology* 174(17): 2880–2896. https://doi.org/10.1111/bph.13912

García-Morales G., M. Huerta-Reyes, M. González-Cortazar, A. Zamilpa, E. Jiménez-Ferrer, R. Silva-García, R. Román-Ramos, and A. Aguilar-Rojas. 2015. "Anti-inflammatory, antioxidant and anti-acetylcholinesterase activities of *Bouvardia ternifolia*: Potential implications in Alzheimer's disease." *Archives of Pharmacal Research* 38(7): 1369–1379. http://doi.org/10.1007/s12272-015-0587-6

Ginwala R., R. Bhavsar, D. I. Chigbu, P. Jain, and Z. K. Khan. 2019. "Potential role of flavonoids in treating chronic inflammatory diseases with a special focus on the anti-inflammatory activity of apigenin." *Antioxidants* 8(2): 35. http://doi.org/10.3390/antiox8020035

Gómez-Aguirre Y. A., A. Zamilpa, M. González-Cortazar, and G. Trejo-Tapia. 2012. "Adventitious root cultures of *Castilleja tenuiflora* Benth. as a source of phenylethanoid glycosides." *Industrial Crops and Products* 36(1): 188–195. http://doi.org/10.1016/j.indcrop.2011.09.005

Graham J. G., M. L. Quinn, D. S. Fabricant, and N. R. Farnsworth. 2000. "Plants used against cancer – An extension of the work of Jonathan Hartwell." *Journal of Ethnopharmacology* 73(3): 347–377. http://doi.org/10.1016/S0378-8741(00)00341-X

Guerrero C., and A. Romo de Vivar. 1973. "Estructura y estereoquímica de la bacchofertina, diterpeno aislado de *Baccharis conferta* H.B.K." *Revista Latinoamericana de Química* 4(4): 178–184.

Gutiérrez-Rebolledo G., M. E. Estrada-Zúñiga, A. Nieto-Trujillo, F. Cruz-Sosa, and A. Jiménez-Arellanes. 2018. "*In vivo* anti-inflammatory activity and acute toxicity of methanolic extracts from wild plant leaves and cell suspension cultures of *Buddleja cordata* Kunth (Buddlejaceae)." *Revista Mexicana de Ingeniería Química* 17(1): 317–330. http://doi.org/10.24275/uam/izt/dcbi/revmexingquim/2018v17n1/Gutierrez

Gutiérrez-Román A. S., G. Trejo-Tapia, M. Herrera-Ruiz, N. Monterrosas-Brisson, J. L. Trejo-Espino, A. Zamilpa, and M. González-Cortazar. 2020. "Effect of terpenoids and flavonoids isolated from *Baccharis conferta* Kunth on TPA-induced ear edema in mice." *Molecules* 25(6): 1379. http://doi.org/10.3390/molecules25061379

Han M. F., X. Zhang, L. Q. Zhang, and Y. M. Li. 2018. "Iridoid and phenylethanol glycosides from *Scrophularia umbrosa* with inhibitory activity on nitric oxide production." *Phytochemistry Letters* 28: 37–41. https://doi.org/10.1016/j.phytol.2018.09.011

Heinrich M., A. Ankli, B. Frei, C. Weimann, and O. Sticher. 1998. "Medicinal plants in Mexico: Healers' consensus and cultural importance." *Social Science and Medicine* 47(11): 1859–1871. http://doi.org/10.1016/s0277-9536

Herrera-Ruiz M., R. López-Rodríguez, G. Trejo-Tapia, B.. E. Domínguez-Mendoza, M. González-Cortazar, J. Tortoriello, and A. Zamilpa. 2015. "A new furofuran lignan diglycoside and other secondary metabolites from the antidepressant extract of *Castilleja tenuiflora* Benth." *Molecules* 20(7): 13127–13143. http://doi.org/10.3390/molecules200713127

Hossain M. S., M. S. Uddin, M. T. Kabir, S. Akhter, S. Goswami, A. A. Mamun, O. Herrera-Calderon, M. Asaduzzaman, and M. M. Abdel-Daim. 2017. "*In vivo* screening for analgesic and anti-inflammatory activities of *Syngonium podophyllum* L.: A remarkable herbal medicine." *Annual Research and Review in Biology* 16(3): 1–12. https://doi.org/10.9734/ARRB/2017/35692

Jain H., N. Dhingra, T. Narsinghani, and R. Sharma. 2016. "Insights into the mechanism of natural terpenoids as NF-κB inhibitors: An overview on their anticancer potential." *Experimental Oncology* 38(3): 68. http://doi.org/10.31768/2312-8852.2016.38(3):158-168

Jiménez M., M. E. Padilla, R. Reyes, L. M. Espinosa, E. Melendez, and A. L. Rocha. 1995. "Iridoid glycoside constituents of *Castilleja tenuiflora*." *Biochemical Systematics and Ecology* 4(23): 455–456.

Jiménez-Arellanes M. A., and M. Z. Pérez-González. 2021. "Modulation of secondary metabolites among Mexican medicinal plants by using elicitors and biotechnology techniques." In *Phenolic Compounds - Chemistry, Synthesis, Diversity, Non-Conventional Industrial, Pharmaceutical and Therapeutic Applications*, edited by F. A. Badria. IntechOpen. https://doi.org/10.5772/intechopen.99888

Jiménez-Suárez V., A. Nieto-Camacho, M. Jiménez-Estrada, and B. Alvarado-Sánchez. 2016. "Anti-inflammatory, free radical scavenging and alpha-glucosidase inhibitory activities of *Hamelia patens* and its chemical constituents." *Pharmaceutical Biology* 54(9): 1822–1830. http://doi.org/10.3109/13880209.2015.1129544

Kang W. S., E. Jung, and J. Kim. 2018. "*Aucuba japonica* extract and aucubin prevent desiccating stress-induced corneal epithelial cell injury and improve tear secretion in a mouse model of dry eye disease." *Molecules* 23(10): 2599. http://doi.org/10.3390/molecules23102599

Kim J. Y., H. J. Lim, D. Y. Lee, J. S. Kim, D. H. Kim, H. J. Lee, H. D. Kim, R. Jeon, and J.-A. Ryu. 2009. "*In vitro* anti-inflammatory activity of lignans isolated from *Magnolia fargesii*." *Bioorganic and Medicinal Chemistry Letters* 19(3): 937–940. http://doi.org/10.1016/j.bmcl.2008.11.103

Kirchhoff P. 1943. *Mesoamerica: Sus Límites Geográficos, Composición Étnica y Caracteres Culturales*, 92–107. Washington, DC: Acta Americana Press.

León-Romero Y., J. L. Trejo-Espino, G. Salcedo-Morales, G. Trejo-Tapia, and S. Evangelista-Lozano. 2019. "Optimización de las condiciones de producción de cápsulas con microestacas para aumentar el tiempo de almacenamiento de la planta medicinal mexicana *Castilleja tenuiflora* (Orobanchaceae)." *Acta Botánica Mexicana* 126. http://doi.org/10.21829/abm126.2019.1442

Leyva-Peralta A. L., G. Salcedo-Morales, V. Medina-Pérez, A. R. López-Laredo, J. L. Trejo-Espino, and G. Trejo-Tapia. 2019. "Morphogenesis and *in vitro* production of caffeoylquinic and caffeic acids in *Baccharis conferta* Kunth." *In Vitro Cellular and Developmental Biology - Plant* 55: 581–589. http://doi.org/10.1007/s11627-019-09977-3

López-Caamal A., and R. Reyes-Chilpa. 2021. "The new world bays (*Litsea*, Lauraceae). A botanical, chemical, pharmacological and ecological review in relation to their traditional and potential applications as phytomedicines." *The Botanical Review* 1–29. https://doi.org/10.1007/s12229-021-09265-z

López-Laredo A. R., Y. A. Gómez-Aguirre, V. Medina-Pérez, G. Salcedo-Morales, G. Sepúlveda-Jiménez, and G. Trejo-Tapia. 2012. "Variation in antioxidant properties and phenolics concentration in different organs of wild growing and greenhouse cultivated *Castilleja tenuiflora* Benth." *Acta Physiologiae Plantarum* 34(6): 2435–2442. http://doi.org/10.1007/s11738-012-1025-8

López-Rodríguez R., M. Herrera-Ruiz, G. Trejo-Tapia, B. E. Domínguez-Mendoza, M. González-Cortazar, and A. Zamilpa. 2019. "*In vivo* gastroprotective and antidepressant effects of iridoids, verbascoside and tenuifloroside from *Castilleja tenuiflora* Benth." *Molecules* 24(7): 1292. http://doi.org/10.1007/s11738-012-1025-8

Maleki S. J., J. F. Crespo, and B. Cabanillas. 2019. "Anti-inflammatory effects of flavonoids." *Food Chemistry* 299. https://doi.org/10.1016/j.foodchem.2019.125124

Malik J., J. Tauchen, P. Landa, Z. Kutil, P. Marsik, P. Kloucek, J. Havlik, and L. Kokoska. 2017. "*In Vitro* antiinflammatory and antioxidant potential of root extracts from ranunculaceae species." *South African Journal of Botany* 109: 128–137. https://doi.org/10.1016/j.sajb.2016.12.008

Martínez M. 1944. *Catálogo de Nombres Vulgares y Científicos de Plantas Mexicanas*. Universidad de Michigan: Fondo de Cultura Económica Press.

Martinez R. M., M. S. Hohmann, D. T. Longhi-Balbinot, A. C. Zarpelon, M. M. Baracat, S. R. Georgetti, F. T. M. C. Vicentini, R. C. Sassonia, W. A. Verri Jr. and R. Casagrande. 2020. "Analgesic activity and mechanism of action of a *Beta vulgaris* dye enriched in betalains in inflammatory models in mice." *Inflammopharmacology* 28: 1663–1675. https://doi.org/10.1007/s10787-020-00689-4

Martínez-Bonfil B. P., G. Salcedo-Morales, A. R. López-Laredo, E. Ventura-Zapata, S. Evangelista-Lozano, and G. Trejo-Tapia. 2011. "Shoot regeneration and determination of iridoid levels in the medicinal plant *Castilleja tenuiflora* Benth." *Plant Cell Tissue and Organ Culture* 107: 195. http://doi.org/10.1007/s11240-011-9970-2

Mata S. 2009. "Atlas de las plantas de la medicina tradicional mexicana." *Universidad Nacional Autónoma de México.* Accessed 20 March 2019. http://www.medicinatradicionalmexicana.unam.mx/atlas.php

Matos-Moctezuma, E. 1994. *Mesoamerica. Vol. 1 of Historía Antigua de México: El México Antiguo, sus Áreas Culturales, los Orígenes y el Horizonte Preclásico,* 49–74. Mexico: Miguel Angel Porrúa Press.

Medina-Pérez V., A. R. López-Laredo, G. Sepúlveda-Jiménez, A. Zamilpa, and G. Trejo-Tapia. 2015. "Nitrogen deficiency stimulates biosynthesis of bioactive phenylethanoid glycosides in the medicinal plant *Castilleja tenuiflora* Benth." *Acta Physiologiae Plantarum* 37(5): 93. http://doi.org/10.1007/s11738-015-1841-8

Mollaei M., A. Abbasi, Z. M. Hassan, and N. Pakravan. 2020. "The intrinsic and extrinsic elements regulating inflammation." *Life Sciences*, 118258. http://doi.org/10.1016/j.lfs.2020.118258

Monja-Mio K. M., D. Olvera-Casanova, M. A. Herrera-Alamillo, F. L. Sánchez-Teller, and M. L. Robert. 2021. "Comparison of conventional and temporary immersion systems on micropropagation (multiplication phase) of *Agave angustifolia* Haw. "Bacanora"." *Biotech* 11: 77. https://doi.org/10.1007/s13205-020-02604-8

Monroy-Ortíz C., and P. Castillo-España. 2007. *Plantas Medicinales Utilizadas en el Estado de Morelos*. 2nd ed. Morelos: Universidad Autónoma del Estado de Morelos Press.

Montes-Hernández E., E. Sandoval-Zapotitla, K. Bermúdez-Torres, and G. Trejo-Tapia. 2015. "Potential hosts of *Castilleja tenuiflora* (Orobanchaceae) and characterization of its haustoria." *Flora-Morphology, Distribution, Functional Ecology of Plants* 214: 11–16. http://doi.org/10.1016/j.flora.2015.05.003.

Montes-Hernández E., E. Sandoval-Zapotitla, K. Bermúdez-Torres, J. L. Trejo-Espino, and G. Trejo-Tapia. 2019. "Hemiparasitic interaction between *Castilleja tenuiflora* (Orobanchaceae) and *Baccharis conferta* (Asteraceae): Haustorium anatomy and C- and N-fluxes." *Botanical Sciences* 97(2): 192–201. https://doi.org/10.17129/botsci.2100

Moreno-Anzúrez N. E., S. Marquina, L. Alvarez, A. Zamilpa, P. Castillo-España, I. Perea-Arango, P. N. Torres, M. Herrera-Ruiz, E. R. D. García, J. T. García, and J. Arellano-García. 2017. "Cytotoxic and anti-inflammatory campesterol derivative from genetically transformed hairy roots of *Lopezia racemosa* Cav. (Onagraceae)." *Molecules* 22(1): 118. http://doi.org/10.3390/molecules22010118

Moreno-Escobar J. A., S. Bazaldúa, M. L. Villarreal, J. R. Bonilla-Barbosa, S. Mendoza, and V. Rodríguez-López. 2011. "Cytotoxic and antioxidant activities of selected Lamiales species from Mexico." *Pharmaceutical Biology* 49(12): 1243–1248. http://doi.org/10.3109/13880209.2011.589454

Netea M. G., F. Balkwill, M. Chonchol, F. Cominelli, M. Y. Donath, E. J. Giamarellos-Bourboulis, D. Golendock, M. S. Gresnigt, M. T. Heneka, H. M. Hoffman, and C. A. Dinarello. 2017. "A guiding map for inflammation." *Nature Immunology* 18(8): 826–831. https://doi.org/10.1038/ni.3790

Niazian M. 2019. "Application of genetics and biotechnology for improving medicinal plants." *Planta* 249: 953–973. http://doi.org/10.1007/s00425-019-03099-1

Nicasio-Torres M., J. Serrano-Román, J. Pérez-Hernández, E. Jiménez-Ferrer, M. Herrera-Ruiz. 2017. "Effect of dichloromethane-methanol extract and tomentin obtained from *Sphaeralcea angustifolia* cell suspensions in a model of kaolin/carrageenan-induced arthritis." *Planta Medica International Open* 4(1): e35–e42. https://doi.org/10.1055/s-0043-108760

Olajide O. A., and S. D. Sarker. 2020. *Anti-Inflammatory Natural Products*. 1st ed. Elsevier Inc., Vol. 55. https://doi.org/10.1016/bs.armc.2020.02.002

Pan M. H., C. S. Lai, and C. T. Ho. 2010. "Anti-inflammatory activity of natural dietary flavonoids." *Food Functions* 1(1): 15. https://doi.org/10.1039/c0fo00103a

Park K. S., B. H. Kim, and Il-M. Chang. 2010. "Inhibitory Potencies of Several Iridoids on Cyclooxygenase-1, Cyclooxygnase-2 Enzymes Activities, Tumor Necrosis factor-α and Nitric Oxide Production *In Vitro*". *Evidence-Based Complementary and Alternative Medicine* 7(1): 41–45. https://doi/org/10.1093/ecam/nem129

Pérez-Hernández J., M. D. P. Nicasio-Torres, L. G. Sarmiento-López, and M. Rodríguez-Monroy. 2019. "Production of anti-inflammatory compounds in *Sphaeralcea angustifolia* cell suspension cultivated in stirred tank bioreactor." *Engineering in Life Sciences* 19(3): 196–205. https://doi.org/0.1002/elsc.201800134

Pesce M., S. Franceschelli, A. Ferrone, M. A. De Lutiis, A. Patruno, A. Grilli, M. Felaco, and L. Speranza. 2015. "Verbascoside down-regulates some pro-inflammatory signal transduction pathways by increasing the activity of tyrosine phosphatase SHP-1 in the U937 cell line." *Journal of Cellular and Molecular Medicine* 19(7): 1548–1556. http://doi.org/10.1111/jcmm.12524

Polturak G., D. Breitel, N. Grossman, A. Sarrion-Perdigones, E. Weithorn, M. Pliner, D. Orzaez, A. Granell, I. Rogachev, and A. Aharoni. 2016. "Elucidation of the first committed step in betalain biosynthesis enables the heterologous engineering of betalain pigments in plants." *New Phytologyst* 210: 269–283. http://doi.org/10.1111/nph.13796

Polturak G., U. Heinig, N. Grossman, M. Battat, D. Leshkowitz, S. Malitsky, I. Rogachev, and A. Aharoni. 2018. "Transcriptome and metabolic profiling provides insights into betalain biosynthesis and evolution in *Mirabilis jalapa*." *Molecular Plant* 11(1): 189–204. http://doi.org/10.1016/j.molp.2017.12.002

Qiu Y. L., X. N. Cheng, F. Bai, L. Y. Fang, H. Z. Hu, and D. Q. Sun. 2018. "Aucubin protects against lipopolysaccharide-induced acute pulmonary injury through regulating Nrf2 and AMPK pathways." *Biomedicine and Pharmacotherapy* 106: 192–199. https://doi.org/10.1016/j.biopha.2018.05.070

Rai A., K. Saito, and M. Yamazaki. 2017. "Integrated omics analysis of specialized metabolism in medicinal plants." *The Plant Journal* 90(4): 764–787. http://doi.org/10.1111/tpj.13485

Rao G. B. S., K. Srisailam, V. U. M. Rao, and B. Vasduda. 2019. "Evaluation of *Galphimia glauca* stem methanol extract fractions for analgesic and anti-inflammatory activities." *Asian Journal of Pharmaceutical and Clinical Research* 12(4): 266–272. https://doi.org/10.22159/ajpcr.2019.v12i4.31866

Reyes-Pérez R., M. Herrera-Ruiz, M., I. Perea-Arango, F. Martínez-Morales, J. J. Arellano-García, and M. D. P. Nicasio-Torres. 2021. "Anti-inflammatory compounds produced in hairy roots culture of *Sphaeralcea angustifolia*." *Plant Cell, Tissue and Organ Culture*. https://doi.org/10.1007/s11240-021-02162-8

Romero-Estrada A., A. Maldonado-Magaña, J. González-Christen, S. Marquina-Bahena, M. L. Garduño-Ramírez, V. Rodríguez-López, and L. Alvarez. 2016. "Anti-inflammatory and antioxidative effects of six pentacyclic triterpenes isolated from the mexican copal resin of *Bursera copallifera*." *BMC Complementary Alternative Medicine* 16(1): 422. http://doi.org/10.1186/s12906-016-1397-1

Roychowdhury D., M. Halder, and S. Jha. 2016. "*Agrobacterium rhizogenes* mediated transformation in medicinal plants: Genetic stability in long-term culture." In *Transgenesis and Secondary Metabolism*, edited by S. Jha, 323–345. Reference Series in Phytochemistry. Cham: Springer Press. http://doi.org/10.1007/978-3-319-27490-4_8-1

Rubio-Rodríguez E., A. R. López-Laredo, V. Medina-Pérez, G. Trejo-Tapia, and J. L. Trejo-Espino. 2019. "Influence of spermine and nitrogen deficiency on growth and secondary metabolites accumulation in *Castilleja tenuiflora* Benth. cultured in a RITA® temporary immersion system." *Engineering in Life Sciences* 19(12): 944–954. http://doi.org/10.1002/elsc.201900040

Rubio-Rodríguez E, I. Vera-Reyes, E. B. Sepúlveda-García, A. C. Ramos-Valdivia, and G. Trejo-Tapia. 2021. "Secondary metabolite production and related biosynthetic genes expression in response to methyl jasmonate in *Castilleja tenuiflora* Benth. *in vitro* plants." *Plant Cell, Tissue and Organ Culture* 114: 519–532. http://doi.org/10.1007/s11240-020-01975-3

Salcedo-Morales G., A. G. Jimenez-Aparicio, F. Cruz-Sosa, and G. Trejo-Tapia. 2014. "Anatomical and histochemical characterization of *in vitro* haustorium from roots of *Castilleja tenuiflora*." *Biologia Plantarum* 58(1): 164–168. http://doi.org/10.1007/s10535-013-0369-2

Salcedo-Morales G., J. L. Trejo-Espino, B. P. Martínez-Bonfil, F. Cruz-Sosa, and G. Trejo-Tapia. 2017. "Formación de raíces e inducción de haustorios de *Castelleja tenuiflora* Benth. con catequina y peróxido de hidrógeno" *Polibotánica* 44: 147–156. http://doi.org/10.18387/polibotanica.44.11

Sanchez P. M., M. L. Villarreal, M. Herrera-Ruiz, A. Zamilpa, E. Jiménez-Ferrer, and G. Trejo-Tapia. 2013. "*In vivo* anti-inflammatory and anti-ulcerogenic activities of extracts from wild growing and *in vitro* plants of *Castilleja tenuiflora* Benth. (Orobanchaceae)." *Journal of Ethnopharmacology* 150(3): 1032–1037. http://doi.org/10.1016/j.jep.2013.10.002

Sarkar J., A. Misra, and N. Banerjee. 2020. "Genetic transfection, hairy root induction and solasodine accumulation in elicited hairy root clone of *Solanum erianthum* D. Don." *Journal of Biotechnology* 323: 238–245. http://doi.org/10.1016/j.jbiotec.2020.09.002

Seo B.-B., Y. Kwon, J. Kim, K. H. Hong, S.-E. Kim, H.-R. Song, Y.-M. Kim, and S.-C. Song. 2021. "Injectable polymeric nanoparticle hydrogel system for long-term anti-inflammatory effect to treat osteoarthritis." *Bioactive Materials* 7: 14–25. https://doi.org/10.1016/j.bioactmat.2021.05.028

Sharma A., P. I. Angulo-Bejarano, A. Madariaga-Navarrete, G. Oza, H. Iqbal, A. Cardoso-Taketa, and M. L. Villarreal. 2018. "Multidisciplinary investigations on *Galphimia glauca*: A Mexican medicinal plant with pharmacological potential." *Molecules* 23(11): 2985. https://doi.org/10.3390/molecules23112985

Shi Q., J. Cao, L. Fang, H. Zhao, Z. Liu, J. Ran, X. Zheng, X. Li, Y. Zhou, D. Ge, H. Zhang, L. Wang, Y. Ran, and J. Fu. 2014. "Geniposide suppresses

LPS-induced nitric oxide, PGE2 and inflammatory cytokine by down-regulating NF-κB, MAPK and AP-1 signaling pathwaysin macrophages". *International Immunopharmacology* 20: 298–306. https://doi.org/10.1016/j.intimp.2014.04.004

Shi M., P. Liao, S. H. Nile, M. I. Georgiev, and G. Kai. 2021. "Biotechnological exploration of transformed root culture for value-added products." *Trends in Biotechnology* 39(2): 137–149. https://doi.org/10.1016/j.tibtech.2020.06.012

Shin M.-S., J. Y. Park, J. Lee, H. H. Yoo, D.-H. Hahm, S. C. Lee, S. Lee, G. S. Hwang, K. Jung, and K. S. Kang. 2017. "Anti-inflammatory effects and corresponding mechanisms of cirsimaritin extracted from *Cirsium japonicum* var. maackii Maxim." *Bioorganic and Medicinal Chemistry Letters* 27(14): 3076–3080. http://doi.org/10.1016/j.bmcl.2017.05.051

Tian X. Y., M. X. Li, T. Lin, Y. Qiu, Y. T. Zhu, X. L. Li, X. W. D. Tao, P. Wang, X. X. Ren, and L. P. Chen. 2021. "A review on the structure and pharmacological activity of phenylethanoid glycosides." *European Journal of Medicinal Chemistry* 209: 112563. https://doi.org/10.1016/j.ejmech.2020.112563

Trejo-Tapia G., G. Rosas-Romero, A. R. López-Laredo, K. Bermúdez-Torres, and A. Zamilpa. 2012. "*In vitro* organ cultures of the cancer herb *Castilleja tenuiflora* Benth. as potential sources of iridoids and antioxidant compounds." *Biotechnological Production of Plant Secondary Metabolites* 87. http://doi.org/0.2174/978160805114411201010087

Trejo-Tapia G., Y. León-Romero, E. B. Montoya-Medina, A. R. López-Laredo, and J. L. Trejo-Espino. 2019. "Perspectives of synthetic seed technology for conservation and mass propagation of the medicinal plant *Castilleja tenuiflora* Benth." In *Synthetic Seeds*. Springer International Publishing. https://doi.org/10.1007/978-3-030-24631-0_15

Udomruk S., C. Kaewmool, P. Pothacharoen, T. Phitak, and P. Kongtawelert. 2018. "Sesamin suppresses LPS-induced microglial activation via regulation of TLR4 expression." *Journal of Functional Foods* 49: 32–43. http://doi.org/10.1016/j.jff.2018.08.020

Valdez-Tapia R., J. Capataz-Tafur, A. R. López-Laredo, J. L. Trejo-Espino, and G. Trejo-Tapia. 2014. "Effect of immersion cycles on growth, phenolics content, and antioxidant properties of *Castilleja tenuiflora* shoots." *In Vitro Cellular and Developmental Biology-Plant* 50: 471–477. http://doi.org/10.1007/s11627-014-9621-5

Wang S. N., G. P. Xie, C. H. Qin, Y. R. Chen, K. R. Zhang, X. Li, Q. Wu, W. Dong, J. Yang, and B. Yu. 2015. "Aucubin prevents interleukin-1 beta induced inflammation and cartilage matrix degradation via inhibition of NF-κB signaling pathway in rat articular chondrocytes." *International Immunopharmacology* 24(2): 408–415. https://doi.org/10.1016/j.intimp.2014.12.029

Wardyn J. D., A. H. Ponsford, and C. M. Sanderson. 2015. "Dissecting molecular cross-talk between Nrf2 and NF-κB response pathways." *Biochemical Society Transactions* 43(4): 621–626. http://doi.org/10.1042/BST20150014

Weimann C., U. Göransson, U. Pongprayoon-Claeson, P. Claeson, L. Bohlin, H. Rimpler, and M. Heinrich. 2002. "Spasmolytic effects of *Baccharis conferta* and some of its constituents." *Journal of Pharmacy and Pharmacology* 54(1): 99–104. http://doi.org/10.1211/0022357021771797

WHO (World Health Organization). 2005. *National Policy on Traditional Medicine and Regulation of Herbal Medicines. Report of WHO Global Survey.* Accessed 20 February 2020. https://apps.who.int/iris/handle/10665/43229

WHO (World Health Organization). 2013. *Traditional Medicine: Executive Board 134th Session.* Accessed 20 February 2020. https://apps.who.int/gb/ebwha/pdf_files/EB134/B134_24-en.pdf

Xian H., F. Weineng, and J. Zhang. 2019. "Schizandrin A enhances the efficacy of gefitinib by suppressing IKKβ/NF-κB signaling in non-small cell lung cancer." *European Journal of Pharmacology* 855: 10–19. http://doi.org/10.1016/j.ejphar.2019.04.016

Yang B., W. Han, H. Han, Y. Liu, W. Guan, and H. Kuang. 2019. "Lignans from *Schisandra chinensis* rattan stems suppresses primary Aβ1-42-induced microglia activation via NF-κB/MAPK signaling pathway." *Natural Product Research* 33(18): 2726–2729. http://doi.org/10.1080/14786419.2018.1466128

Yang H., Z. Du, W. Wang, M. Song, K. Sanidad, E. Sukamtoh, J. Zheng, L. Tian, H. Xiao, Z. Liu, and G. Zhang. 2017. "Structure–activity relationship of curcumin: Role of the methoxy group in anti-inflammatory and anticolitis effects of curcumin." *Journal of Agricultural and Food Chemistry* 65(22): 4509–4515. http://doi.org/10.1021/acs.jafc.7b01792

Yoou M. S., H. M. Kim, and H. J. Jeong. 2015. "Acteoside attenuates TSLP-induced mast cell proliferation via down-regulating MDM2." *International Immunopharmacology* 26(1): 23–29. https://doi.org/10.1016/j.intimp.2015.03.003

Yoshida S., S. Cui, Y. Ichihashi, and K. Shirasu. 2016. "The haustorium, a specialized invasive organ in parasitic plants." *Annual Review Plant Biology* 29(67): 643–667. http://doi.org/10.1146/annurev-arplant-043015-111702

Zeng X., F. Guo, and D. Ouyang. 2020. "A review of the pharmacology and toxicology of aucubin." *Fitoterapia* 140: 104443. http://doi.org/10.1016/j.fitote.2019.104443

Zhao Y.-L., X.-W. Yang, B.-F. Wu, J.-H. Shang, Y.-P. Liu, ZhiDai, and X.-D. Luo. 2019. "Anti-inflammatory effect of pomelo peel and its bioactive coumarins." *Journal of Agricultural and Food Chemistry* 67(32): 8810–8818. https://doi.org/10.1021/acs.jafc.9b02511

Zhou Y., M. Lihui, S. Yunxia, W. Mengying, and F. Xiang. 2020. "Pharmacodynamic effects and molecular mechanisms of lignans from *Schisandra chinensis* Turcz. (Baill.), a current review." *European Journal of Pharmacology* 173796. http://doi.org/10.1016/j.ejphar.2020.173796

CHAPTER 3

Production of Secondary Metabolites by Hairy Root Cultures

Alma Angélica Del Villar-Martínez,
Edmundo Lozoya-Gloria, and
Pablo Emilio Vanegas-Espinoza

CONTENTS

DOI: 10.1201/9781003166535-3

3.1 INTRODUCTION

Agrobacterium is a Gram-negative bacterial genus well-known for being able to transfer part of its DNA to plant genomes during a natural infection process leading to tumors described as crown galls or abnormal roots named hairy roots (Chen and Otten 2017).

Hairy roots are induced by *A. rhizogenes* which transfers T-DNA carried in the root-inducing (Ri) plasmid into the genome of the infected plant (Dessaux and Faure 2018; Stavnstrup et al. 2020). As a result, hairy roots are formed at the infection site where genetic transformation occurs (Otten 2021). The *rol* genes contribute towards the abnormal growth of hairy roots, as well as the production and accumulation of bioactive compounds (Dhiman et al. 2018). The hairy roots are characterized by their genetic and biosynthetic stability, rapid and exogenous-hormone-independent growth, lateral branching, and absence of geotropism (Gutierrez-Valdes et al. 2020). This chapter shows the achievements in the use of hairy roots generated by *A. rhizogenes* to produce secondary metabolites of pharmacological interest, taking into consideration all advantages offered by this biotechnological process.

3.2 INDUCTION AND ESTABLISHMENT OF HAIRY ROOT CULTURE BY *AGROBACTERIUM RHIZOGENES* INFECTION

A. rhizogenes is a microorganism that lives in the rhizospheric soil, first identified as *Rhizobium rhizogenes* in the 1930s and given its current name in 1942. This bacterium induces the hairy root syndrome, characterized by root overgrowth on plant tissue at the infection site. *Agrobacterium* spp. are ubiquitous Gram-negative soil bacteria with a cosmopolitan distribution; pathogenic species such as *Agrobacterium tumefaciens* and *A. rhizogenes* cause crown gall and hairy root diseases, respectively. These bacteria have long been considered a problem in agriculture. Nevertheless, in many research laboratories, agrobacteria are associated not with plant disease, but rather with their innate ability to transfer DNA into cells of many plant species (Saeger et al. 2020). As mentioned before, following the infection–transformation process with *A. rhizogenes*, numerous and different transformed hairy root lines can be generated.

A. rhizogenes strains carry the Ri (root-inducing) plasmid which has a region known as T-DNA which contains genes involved in the initiation and development of roots, namely *rol* genes, which are also essential for opine biosynthesis (Matveeva and Otten 2021). Several studies have revealed that only four open reading frames (ORFs) of the T-DNA are critical for the induction, growth, and morphology of hairy roots in infected plants. These *loci* are called *rol* oncogenes or root-inducing *loci*, including: *rolA* (ORF10), *rolB* (ORF11), *rolC* (ORF12), and *rolD* (ORF15) (Sarkar et al. 2018). Depending on the type of Ri plasmid that the specific *A. rhizogenes* strain contains, they are grouped into agropine, mannopine, octopine, or cucumopine types (Table 3.1).

The agropine-type strain contains the Ri plasmid in which its T-DNA is truncated as split T-DNA, symbolized as TL-DNA (left) and TR-DNA

TABLE 3.1 Different Strains Used in Hairy Roots Induction.

Opine Type	Strain	Transgenes Detected by PCR	Plant Species	Reference
Agropine	A4	*rolA, rolB, rolC, rolD, aux1, aux2 y virG*	*Rhaponticum carthamoides*	Skała et al. 2015
		rolC y virD2	*Bonellia macrocarpa*	Ruiz-Ramírez et al. 2018
		rolA, rolB, rolC	*Solanum tuberosum*	Moehninsi and Navarre 2018
		rolB	*Origanum vulgare*	Habibi et al. 2016
		rolC, aux1, virD2	*Camellia sinensis*	Rana et al. 2016
		rolA, rolB, rolC, rolD, ags, virD y hptII	*Lotus corniculatus*	Savic et al. 2019
		rolA, rolB, rolC, rolD, aux1, aux2, virG	*Rhaponticum carthamoides*	Skała et al. 2015
	R1000	*rolB*	*Withania somnifera*	Thilip et al. 2015
		rolB, rolC	*Raphanus sativus*	Muthusamy and Shanmugam 2020
	MTCC 532	*rolB, rolC*	*Berberis aristata*	Brijwal and Tamta 2015
Cucumopine	2659/K599	*rolB*	*Origanum vulgare*	Habibi et al. 2016
Mannopine	8196	*rolB, ORF13*	*Solanum lycopersicum*	Lima et al. 2009
Mikimopine	1724	*rolA, rolB, rolC, ORF13, ORF14*	*Ajuga reptans*	Tanaka, Yamakawa and Yamashita 1998

(right). Whereas the TL-DNA segment has been found to contain the hairy root-inducing root oncogenic *loci* (*rol*), the TR-DNA has been found to contain the genes for auxin and opine biosynthesis (Singh et al. 2018). It has been reported that T-DNA can be truncated at the left and/or right ends before being inserted into the plant genome, which complicates the identification of T-DNA inserted (Sun et al. 2019). It has also been documented that, although TL-DNA and TR-DNA are independently transferred in the plant genome, the integration of TL-DNA in the plant genome is essential for hairy root induction (Singh et al. 2018).

Opines are secondary amine derivatives of low molecular weight compounds, formed by condensation of amino acids, either with a keto acid or a sugar, providing a favorable environment to the growth of the bacteria (Matveeva and Otten 2021). There are two subcategories of opines: the first includes the opines from the nopaline (nopaline, nopalinic acid, leucinopine, glutaminopine, succinamopine) and octopine/cucumopine (octopine, octopinic acid, lysopine, histopine) families; the second subcategory includes the mannityl (mannopine, mannopinic acid, agropine, agropinic acid) and the chrysopine (deoxy-fructosyl glutamate, deoxy-fructosyl glutamine, deoxy-fructosyl oxoproline, and chrysopine) families (Padilla et al. 2021).

Gene insertion has been detected efficiently by PCR amplification of *rol* genes. In addition, in some investigations, reporter genes have been used, specifically strains of *A. rhizogenes* such as K599 harboring the *p35SGFPGUS+* plasmid. This binary vector harbors two reporter genes: *uidA* gene (*GUS*, encoding β-glucuronidase) fused to a catalase intron, followed by the *GFP* (green fluorescent protein) gene; both genes were under the control of cauliflower mosaic virus 35S (*CaMV35S*) constitutive promoter (Garagounis et al. 2020). The A4 *A. rhizogenes* strain contains the binary plasmid *pBI121*, which comprises the *uidA* reporter gene driven by the *CaMV35S* promoter and with the nopaline synthase terminator (*nos*) gene (Alagarsamy et al. 2018). The use of the tomato threonine deaminase (*TDT*) protein (Ruiz-Ramírez et al. 2018; Bazaldúa et al. 2019), *pGus-GFP+* (Bazaldúa et al. 2019), *GUS*, *GFP*, and *DsRed2* genes (Chen and Otten 2017) have also been efficiently used as reporter proteins to detect the transformation mediated by *A. rhizogenes*.

The β-glucuronidase (*GUS*) reporter gene is used to detect the insertion of DNA into the plant genome using either of the methods of DNA transfer, such as biolistic bombardment or *Agrobacterium*-mediated transformation. This assay is destructive, leaving no chance for proliferation and regeneration of identified transformants; this is a major disadvantage of using the *GUS* reporter gene (Balakrishna et al. 2019).

Green fluorescent protein (GFP) from the *Aequorea victoria* jellyfish is nowadays the most widely used and developed reporter used in biochemistry and cell biology (Wons et al. 2018). GFP gained importance because of its non-destructive visualization systems that can facilitate the recovery of identified transformed tissues (Balakrishna et al. 2019).

Hairy root culture is a promising biotechnological approach that could be implemented to preserve rare, valuable, threatened, or endemic medicinal species, to protect the biodiversity of such species, and, for the *in vitro* production of valuable secondary metabolites (Gutierrez-Valdes et al. 2020).

The quality and quantity of metabolites extracted from hairy root cultures are similar to those found in wild-type plants, although it is possible to improve the chemical profile regarding the interesting compounds; moreover, these cultures could accumulate novel secondary metabolites not detected in non-transformed tissues (Otten 2018; Singh et al. 2018; Gutierrez-Valdes et al. 2020). In general, to establish the culture of transformed roots by *A. rhizogenes*, five parameters must be considered (Figure 3.1):

1. Selection of an *A. rhizogenes* strain with the ability to infect the plant species in question;
2. Selection of an explant that is susceptible to being infected by the agrobacteria;
3. Availability of an appropriate infection method by which the bacteria can efficiently infect the plant tissue;
4. Induction of hairy roots at the site of infection;
5. Establishment of root cultures in a medium without exogenous plant growth regulators.

There is a wide range of plants that have been successfully infected by *A. rhizogenes*, as well as suitable explants capable of being infected (Table 3.2). Hairy roots induction does not depend only on the genotype of the plant, but also on the bacterial strain, the infection method, and the culture conditions, among other factors (Singh et al. 2018; Kanchanamala and Bandaranayake 2019).

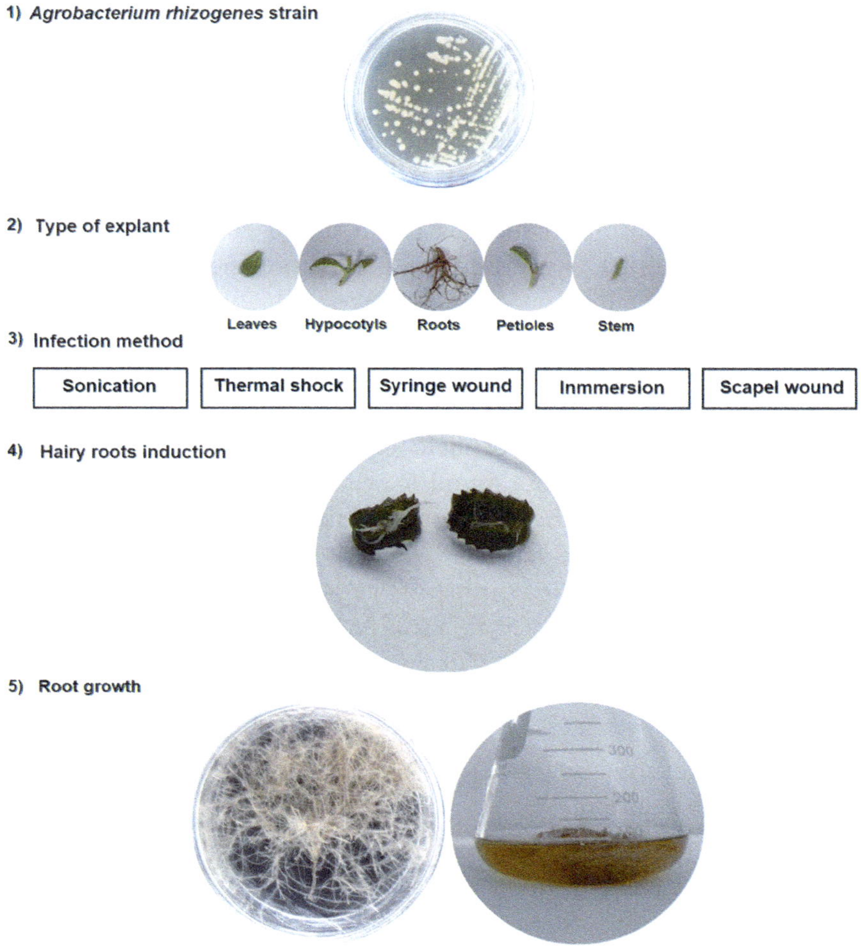

FIGURE 3.1 Induction process and establishment of hairy roots in hormone-free medium.

In general, there are four main steps of T-DNA transfer from *Agrobacterium* to the plant cell genome:

1) The induction of expression of *vir* genes and the generation of the single-stranded T-DNA: *Agrobacterium* perceives many plant-derived chemicals, that includes wound-releasing chemicals (Guo et al. 2019) that induce *Agrobacterium vir* genes expression.

TABLE 3.2 Hairy Roots Induction by *Agrobacterium rhizogenes* Strains on Different Explants and in Different Plant Species.

Plant	Explant	Strains	References
Tapa natans	Stem and leaf nodes	15834 and K599	Mikhaylova et al. 2021
Ardisia crenata	Leaves	ATCC 15834 and A4	Hu et al. 2020
Atropa komarovii	Leaves	ATCC 15834	Banihashemin et al. 2020
Cucumis anguria	Leaves, cotyledons, hypocotyls, and shoot nodes	LBA9402, A4, 15834, 13333, R1200, R1000, R1301 and R1601	Sahayarayan et al. 2020
Lithospermum erythrorhizon	Leaves and stems	A13 harboring pBI121_GFPh	Tatsumi et al. 2020
Rhodiola rosea	Leaves, stems, and rhizomes	ATCC43057 harboring pRiA4	Martínez et al. 2020
Salvia bulleyana	Leaves and shoots	A4	Wojciechowska et al. 2020
Selaginella bryopteris	Rhizophore, stem, and fronds	LBA 1334	Singh et al. 2020
Solanum erianthum	Leaves	A4	Sarkar et al. 2020
Trachyspermum ammi	Seedling stem	A4, LBA 9402, ATCC 15834	Vamenani et al. 2020
Trigonella foenum-graecum	Hypocotyl	ATCC15834, R1000, A4 and C58	Zolfaghari et al. 2020
Ferula pseudalliacea	Leaves, hypocotyls, cotyledons, roots	Ar318, ArA4, ATCC 15834, 1724 and LBA9402	Khazaei et al. 2019
Dracocephalum kotschyi	Leaves	ATCC15834	Nourozi et al. 2019
Gentiana urticulosa	Shoots	A4M70GUS	Vinterhalter et al. 2019
Lemna minor	Leaves and root tips	MSU 440 harboring pBIN-YFP	Kanchanamala and Bandaranayake 2019
Camellia sinensis var. *sinensis*	Hypocotyl	A4 and A4 harboring pBI121	Alagarsamy et al. 2018
Cichorium spp.	Leaves and leaf stalks	K599 harboring p35SGFPGUS+ plasmid	Hanafy et al. 2018
Polygonum multiflorum	Leaves	KCCM 11879	Ho et al. 2018

(*Continued*)

TABLE 3.2 (CONTINUED) Hairy Roots Induction by *Agrobacterium rhizogenes* Strains on Different Explants and in Different Plant Species.

Plant	Explant	Strains	References
Macleaya cordata	Leaves and stems	10060	Huang et al. 2018
Trifolium pratense	Leaves, petioles, roots, and shoot tips	MTCC 532 and MTCC 2364	Kumar et al. 2018
Althea officinalis	Leaves, petioles, and shoots	A4, A13, ATCC15834, and $ATCC15834_{(GUS)}$	Tavassoli and Afshar 2018

2) The T-DNA covalently attaches to VirD2, and several *vir*-encoded effector proteins which together are exported out of the bacterial cell (Gelvin 2017).

3) After reaching the cytoplasm, virulence-effector proteins and T-DNA strands target the nucleus.

4) During the integration of T-strands to the host plant genome, both vir-effector proteins and plant proteins are removed from T-strands.

3.3 EFFECT OF THE *rol* GENES OF *AGROBACTERIUM RHIZOGENES* ON THE MORPHOLOGY AND ACCUMULATION OF SECONDARY METABOLITES IN MEDICINAL PLANTS

Differences in hairy root morphology include variation in the thickness of primary roots, the lateral density of the roots, and spontaneous callus generation, among others (Sarkar et al. 2018). Figure 3.2 shows the variable morphology found in different *Kalanchoe* transformation events, such as unbranched poorly haired roots or abundant branched hairiness in comparison with wild-type roots.

Hairy root phenotype can change depending on the individual transformation event, the positioning of the T-DNA integration in the plant genome, the T-DNA copy number, and eventual fragmentation and expression levels of the associated *aux* and *rol* genes (Desmet et al. 2020). Hairy root cultures are characterized by their profuse neoplastic growth on a hormone-free medium, with non-geotropic, branching overgrowth, which can allow the successful production of different secondary metabolites (Gutierrez-Valdes et al. 2020).

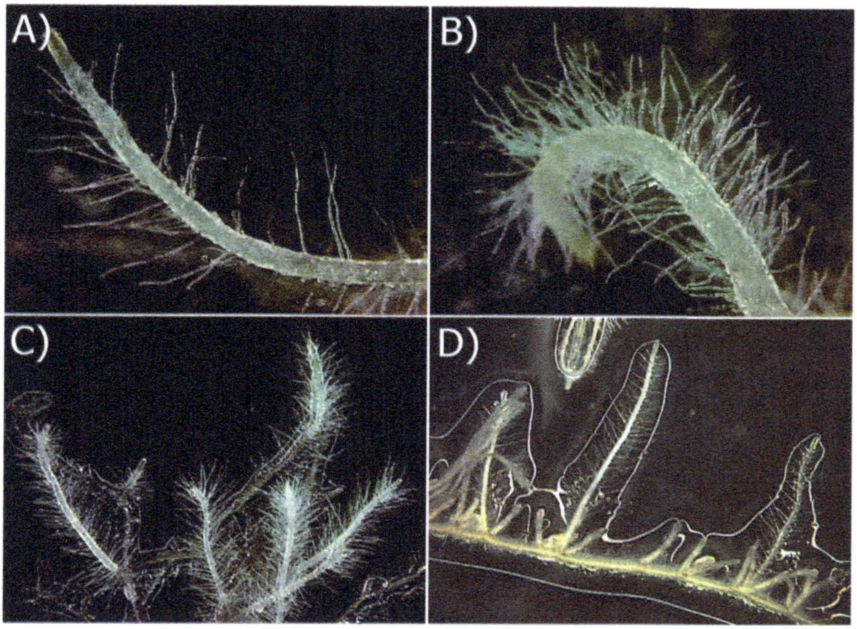

FIGURE 3.2 Morphological appearance of hairy roots of *Kalanchoe* spp. A, B, C) hairy roots D) *in vitro* wild-type roots.

3.4 PLANT REGENERATION FROM HAIRY ROOT CULTURES

Regeneration of plants from Ri-transformed roots has been reported in many plant species; these plants exhibit unique characteristic features, known as the "hairy root syndrome". Ri-transformed plants are morphologically characterized by stunted growth with a reduced shoot and internodes lengths, accompanied by an increase in the number of nodes, and leaves. Hairy root syndrome of Ri-transformed plants is due to the insertion and expression of *rol* genes of the TL-DNA of the Ri plasmid.

The *rolB* and *rolC* genes in the T-DNA are named *plast* genes because of their capacity to change the development of the regenerated transformed plants obtained from hairy roots (Otten 2018). Binary vectors carrying specific gene constructs can be introduced into oncogenic *A. rhizogenes* strains; the resulting bacteria can then be used to obtain co-transformed hairy roots which integrate both the T-DNA from the Ri plasmid and the T-DNA from the genetically engineered binary vector. The co-transformation procedure enables a more rapid analysis of transformed roots than the methods used to generate plants that are stably transformed by disarmed

A. tumefaciens or by direct gene transfer techniques, such as biolistic or protoplast electroporation (Zhong et al. 2018).

Secondary metabolite biosynthesis is not always limited to roots and underground parts; in many plants, the precursors of pharmaceutically important compounds are produced in the roots, but the final products are synthesized in leaves or other aerial parts of the plants (Dhiman et al. 2018; Neumann et al. 2020).

There is an emerging interest in the ornamental plant industry of plants regenerated from hairy root tissue, because of specific morphological characteristics such as dwarfing, increased branching, wrinkled leaves, decreased apical dominance, and enhanced root growth due to the Ri phenotype (Desmet et al. 2020). During *Kalanchoe* hairy roots culture, spontaneous induction of somatic embryos has been observed, which germinate into whole plants; this strategy represents an alternative method of producing plants with pharmacological potential (Figure 3.3).

FIGURE 3.3 Hairy roots culture: spontaneous induction of somatic embryos, and biomass accumulation in plant growth regulator-free medium.

3.5 *AGROBACTERIUM RHIZOGENES* AS A BIOTECHNOLOGICAL TOOL FOR SECONDARY METABOLITE PRODUCTION

The discovery of molecules responsible for the effect of plants used in folk medicine has given a great impulse to the study of new plants and strategies to induce a greater accumulation of bioactive compounds and opens up the possibilities of producing new compounds in plant cells, tissues, and organs. The establishment of transformed root cultures offers the possibility of having a constant source of rapidly growing biomass, independently of plant growth regulators. This is important to produce metabolites that are specifically synthesized and accumulated in the roots. However, it has been shown that the genetic transformation that generates the hairy roots phenotype sometimes causes modifications in the root's metabolism, inducing the synthesis of novel compounds not found in wild-type roots (Otten 2018; Singh et al. 2018; Gutierrez-Valdes et al. 2020; Gantait and Mukherjee 2021). Another advantage of culturing hairy roots is that after the infection–transformation process, each root generated represents an independent transformation event. Then, numerous transformed root lines can be generated, with the possibility that each line may have a modified metabolism.

Nakasha et al. (2017) developed an efficient hairy roots induction protocol for *Chlorophytum borivilianum*. *In vitro* shoots were infected with MAFF106590, MAFF106591, MAFF301726, MAFF210265, and MAFF720002 *A. rhizogenes* strains, but only MAFF720002, MAFF106590, and MAFF106591 induced hairy roots, whereas the other two strains failed to transfer their T-DNA to plant tissues. The *A. rhizogenes* MAFF720002 strain responded positively by transforming all *C. borivilianum* stem sizes. Stems of 2.5 cm in length had the highest transformation percentage (76 to 81%), the highest mean±SE number of hairy roots per explant (5.02±0.06), and the longest hairy roots (23±0.49 cm). PCR analysis detected the *rolB* gene in all transformed samples.

The *A. rhizogenes* strain played a significant role in producing and promoting the growth of hairy roots in *C. borivilianum*. Hairy roots induction, after infection of leaves or leaf stalks of *Cichorium intybus* and *Cichorium endivia*, was reported by Hanafy et al. (2018). The *A. rhizogenes* K599 strain, harboring the *p35SGFPGUS+* plasmid, proved to be competent for the transformation of both species tested. Hairy roots of the two species were isolated, from which eleven *C. intybus* and two *C. endivia* clones were selected due to their fast-growing behavior. Furthermore, liquid MS-basal medium proved to be the most suitable for biomass production, whereas PCR analysis revealed foreign DNA integration in the selected transgenic hairy root clones. Hairy root lines exhibited both higher growth rates and higher inulin accumulation than non-transgenic roots.

Huang et al. (2018) established hairy root cultures of *Macleaya cordata*, the main source of the alkaloid sanguinarine which is widely used to replace growth-promoting antibiotics in animal feed and has shown useful anticancer activity. Hairy roots were obtained by co-cultivation of leaf and stem explants with *A. rhizogenes* 10060 strain. Moreover, the concentrations of sanguinarine and dihydrosanguinarine in hairy roots were much higher than in the wild-type roots. Kumar et al. (2018) reported the development of an efficient technique for large-scale cultivation of fast-growing hairy roots culture systems for bioactive isoflavones production. Hairy roots of *Trifolium pratense* were induced by infection with *A. rhizogenes* strains MTCC 532 and MTCC 2364 strains. Fast-growing hairy root lines of *T. pratense* were evaluated and selected for their growth and isoflavone production. Hairy root line 2364A displayed significantly higher accumulations of all four pharmaceutically important isoflavones, namely daidzein (8.56 mg/g), genistein (2.45 mg/g), formononetin (15.23 mg/g), and biochanin A (1.10 mg/g), compared with other lines. However, it is necessary to be aware that *A. rhizogenes* infection is not successful in all plant species. As mentioned before, this process depends on the bacterial strain, the infected tissue, the culture medium, etc. Hairy root cultures of *Ardisia crenata* were established with strains ATCC 15834 and A4, with ATCC 15834 showing the highest transformation frequency. The ardicrenin concentration in the hairy roots was significantly improved by 8.2%, compared with wild-type roots, and was also 2.7-, 9.4-, and 2.6-fold greater (4.5, 3.1, 0.8 and 3.2%, respectively) than those in non-transformed cultured roots (Hu et al. 2020). Strain LBA1334 of *A. rhizogenes* induced hairy roots in *Selaginella bryopteris* after 48 h co-cultivation and 6 d of infection. PCR amplification with *rol* A and *vir* C gene-specific primers confirmed the Ri T-DNA integration in the induced hairy roots. This could be an efficient, stable, and viable alternative *in vitro* method for the production of bioactive compounds in *S. bryopteris* (Singh et al. 2020).

3.6 HAIRY ROOT CULTURES TO PRODUCE SECONDARY METABOLITES WITH VALUABLE BIOLOGICAL ACTIVITY

Higher plants synthesize a wide range of secondary metabolites, such as alkaloids, flavonoids, quinones, lignans, steroids, and terpenoids, which are used as pharmaceuticals, agrochemicals, flavors, fragrances, colors, biopesticides, and food additives (Thakur et al. 2019). Research on plant secondary metabolites has increased during the past 50 years due to the requirements in daily lives, including health care, for these plant products (Jamwal et al. 2018). Evaluation of biological activities of hairy root extracts in a cytotoxic assay in human cancer cell lines, or for antibacterial, antioxidant, anti-inflammatory, antimicrobial, antifungal, and antiviral activities have been reported.

Chung et al. (2016) induced hairy roots of *Brassica rapa* ssp. Rapa through infection of leaves, hypocotyls, and roots with *A. rhizogenes* KCTC 2703 strain. Concentrations of phenolic and flavonoid compounds, and expression levels of *PAL, CHI*, and *FLS* genes were higher in hairy roots than in non-transformed roots. Moreover, antioxidant and antimicrobial activity against bacterial and fungal species, and inhibition of human breast and colon cancer cell lines were higher in hairy root extracts than in wild-type root extracts. Thiruvengadam et al. (2016) reported that antioxidant, antibacterial, antifungal, and antiviral activities, but not phenolic compound concentrations, were higher from extracts of hairy root cultures of *Momordica dioica* obtained by infection of leaves, petioles, and internodal explants with *A. rhizogenes* KCTC 2703 strain than from extracts of non-transformed roots. Kochan et al. (2020) reported the biological activity of extracts of ginseng hairy roots against the Caco-2 human colorectal adenocarcinoma cell line because it had been reported that American ginseng (*Panax quinquefolium*), of which the main bioactive components are ginsenosides, exhibits a range of anti-inflammatory, hepatoprotective, antidiabetic, antiobesity, antihyperlipidemic, and anticarcinogenic functions. Zhang et al. (2020) reported the profile of secondary metabolites produced by *Tripterygium wilfordii* hairy roots after modification of the culture medium to enhance the concentrations of specific active compounds, such as wilforgine and wilforine. The authors concluded that nitrogen and phosphorus were determinants which could be manipulated to increase the accumulation of these compounds by over 40% in hairy roots. Sahai and Sinha (2021) reported the induction of hairy roots on *Taxus baccata* ssp. *wallichiana* through infection with *A. rhizogenes* MTCC 532 strain. The taxol content in the methanolic hairy root extract was 12 mg/g of hairy root biomass, which represents 79% more than the natural wild-type production of 0.15 mg/g of bark, suggesting a possible alternative for commercial taxol production on a large scale.

In the laboratory, induction, and establishment of hairy roots of a range of plant species have been achieved through deliberate infection and co-cultivation of host plant parts with *A. rhizogenes* (Srivastava et al. 2017). Although several reports have demonstrated that established hairy root cultures produce enhanced amounts of secondary metabolites compared to non-transformed plants, in most species the productivity of specific desired compounds by hairy root cultures is generally too low to fulfill the demands of the pharmaceutical industry, owing to various biological and technological limitations (Pedreño and Almagro 2020).

Accumulation of any secondary metabolites in hairy root cultures is controlled by two-stage events: biomass growth and metabolite accumulation. Biomass accumulation represents the first stage of the final yield of the desired secondary metabolite that is greatly influenced by the parameters that control the growth and multiplication of cultured cells/organs, whereas the second stage is represented by the concentration of the metabolite in cells/organs, which is controlled by the parameters that influence secondary metabolites pathways (Halder et al. 2018).

3.7 BIOTECHNOLOGICAL INTERVENTIONS TO IMPROVE METABOLITE PRODUCTION IN HAIRY ROOTS

Several strategies have been employed to make hairy root-based technology viable to produce therapeutic molecules on a large scale; these include medium optimization, precursor feeding, elicitation, and metabolic engineering (Srivastava et al. 2017). There are two types of elicitors, depending upon their origin, namely abiotic and biotic.

Abiotic elicitors mainly include inorganic compounds and physical factors, while biotic elicitors comprise compounds of biological origin. Biotic elicitors include plant signaling molecules (such as methyl jasmonate (MJ), salicylic acid (SA), and the ethylene-releasing compound ethephon), molecules derived from microorganisms (chitosan, polysaccharide fractions of yeast and bacterial extracts, mycelial extracts, glycans, and culture filtrates of root endophytic fungus) and precursors of important metabolites such as phenylalanine and cysteine (Dhiman et al. 2018). Kochan et al. (2020) reported MJ elicitation of hairy root lines of *P. quinquefolium*, concluding that extracts from elicited lines exhibit a stronger inhibitory effect on cellular metabolic activity of a Caco-2 human adenocarcinoma cell line than non-elicited hairy root lines. In hairy root lines of *Valeriana jatamansi*, different concentrations of three elicitors, i.e. MJ, jasmonic acid (JA), and SA were tested on growth rate and valtrate concentration. Valtrate accumulation was increased by the addition of 100 mg/L MJ (3.63 times higher than the control), whereas SA was not an effective elicitor. These results demonstrated that different biotic elicitors have varied inductive effects, depending on the plant species and the type of bioactive compounds (Zhao and Tang 2020). In order to enhance the production of *trans*-resveratrol, *trans*-arachidin-1, and *trans*-arachidin-3, Eungsuwan et al. (2021) reported the use of an elicitation strategy consisting of a pretreatment with paraquat, followed by MJ, and cyclodextrin to *Arachis hypogaea* hairy root cultures, concluding that elicitation treatment increased antioxidant activity, and antimicrobial effect against *Staphylococcus aureus*, *Salmonella typhimurium*, and *Escherichia coli*. The establishment of *Lepechinia caulescens* hairy root cultures has been reported by Vergara-Martínez et al. (2021); these cultures accumulate pharmaceutically important concentrations of ursolic (UA), oleanolic (OA), and rosmarinic (RA) acids, representing an alternative to producing these bioactive metabolites. Different sucrose concentrations (15, 30, 45, or 60 g/L) in the culture medium were evaluated; hairy roots grown in MS/B5 medium, containing MS salts (Murashige and Skoog 1962), B5 vitamins (Gamborg, Miller and Ojima 1968) supplemented with 45 g/L sucrose, accumulate significantly higher UA, OA, and RA concentrations than the wild plant. The authors also reported that the accumulation of these bioactive metabolites was dependent on the MeJA concentrations, and the duration of exposure to the elicitor. Alsoufi et al. (2021) reported the effect of auxins and cytokinins on hairy root cultures of *Calendula officinalis*. The saponin concentration in hairy root cultures and their release into the culture medium were increased by BAP which might indicate that this phytohormone can be used as an elicitor for this system; the auxins tested showed the opposite effect, and the accumulation and release of saponins decreased, whereas the production of phytosterols was mainly increased by kinetin.

3.8 SYNTHETIC SEEDS FOR HAIRY ROOT LINES STORAGE

Artificial seeds are also known as “synseeds” (Murashige 1977). Synthetic or artificial seeds are encapsulated plant tissues such as shoot buds, axillary buds, somatic embryos, shoot tips, cell aggregates, or any other tissues that can be cultured and grown to generate a complete plant under either *in vitro* or *ex vitro* conditions and have the potential to retain their viability after cold storage (Qahtan et al. 2019). Artificial seeds are also valuable in terms of their role in providing protective coating, increasing the level of the micropropagules’ success in the field. These micropropagules need a protective coating to increase successful establishment in the field, because of the sensitivity of uncovered micropropagules to drought and pathogens under natural environmental conditions. Furthermore, artificial seeds are more durable with respect to handling, transportation, and storage; their production is also a useful technique as a clonal propagation system in terms of the preservation of the genetic uniformity of plants, delivery to the field, low cost, and fast propagation of plants (Rihan et al. 2017). Synthetic seeds are mainly divided into two groups either desiccated or hydrated encapsulated synthetic seeds. The desiccated synthetic seeds are produced by the desiccation of somatic embryos, either naked or encapsulated in polyethylene glycol. Hydrated synthetic seeds are produced by encapsulating somatic embryos or other propagules into hydrogel capsules (Kocak et al. 2019). Gangopadhyay et al. (2011) described the role of different exogenous hormones on the morphology and plumbagin production in *Plumbago indica* hairy roots and encapsulated hairy roots. Hairy roots were treated with a single hormone or a combination of different phytohormones; IAA, IBA, 2, 4-D, NAA, BAP, GA_3, and ABA. Cultures incubated with GA_3 (0.5 mg/L) yielded the highest root growth due to the formation of profuse lateral branching whereas NAA (0.5 mg/L) treatment caused the highest plumbagin accumulation. In a combinatorial study, GA_3 + NAA (0.5 mg/L each) was found to be optimum for root biomass and plumbagin production at an earlier stage of culture. Based on the effect of exogenous hormones, hairy root culture treated with GA_3 was selected and encapsulated with a sodium alginate matrix. Uniform-shaped alginate-coated synthetic seeds were conserved for up to six months and exhibited high regeneration potential without disturbing plumbagin content.

Rawat et al. (2013) reported the first induction of *Picrorhiza kurrooa* hairy roots, through infection with *A. rhizogenes* A4 strain, and focused on genetic fidelity and phytochemical profile of conserved tissues following their conversion from encapsulated hairy roots. Among five *P. kurrooa* hairy root lines established, hairy root line H7 was selected for encapsulation due to its high accumulation of picrotin and picrotoxinin. The phytochemical analysis also showed that picrotin and picrotoxinin concentrations (8.3 and 47.6 µg/g DW, respectively) were similar in the hairy root line and its regenerates. Re-growth of encapsulated roots induced the formation of adventitious shoots with 73% frequency on MS medium supplemented with 0.1 mM 6-benzylaminopurine (BAP), following six months of storage at 25 °C. After two months, 85% of regenerated plantlets survived. Genetic fidelity analysis of transformed plants, using

RAPD or ISSR, revealed 5.2% or 3.6% polymorphism, respectively. The phytochemical analysis also showed that picrotin and picrotoxinin concentrations were similar in the hairy root line and its regenerants.

3.9 LARGE-SCALE PRODUCTION OF HAIRY ROOTS.

Hairy roots grow in a suitable medium without any plant growth regulator. The aim of hairy root cultures is the production of secondary metabolites at an industrial level (Rawat et al. 2019). Bioreactors for plant cells and hairy roots have operating conditions like microbial bioreactors, with modified features to achieve efficient growth. Plant cell cultures require aerobic bioreactors with low shear and good mixing. The plant cells are bigger than microbial cells and form aggregates, either cell or organ cultures in suspension, like hairy roots, which makes the sampling of biomass from the bioreactor at constant intervals rather difficult. Measurement of the medium conductivity is an indirect way of estimating the biomass growth in the bioreactor (Srikantan and Srivastava 2018). Normally, a typical scale-up that starts in a laboratory includes jars or shake flasks of 50–250 mL volume, which further rises to a 500 mL–10 L small-scale bioreactor. However, for large or industrial scale, stainless steel vessels of varying sizes (>10 L) are required. Several physical and chemical aspects need to be minutely attended to during the scaling-up of hairy root cultures (Mehrotra et al. 2018). Hairy root bioreactors can generally be divided into gas or liquid phases. In liquid-phase bioreactors, roots are always placed in the medium; as a result, they are called submerged reactors. On the other hand, in gas-phase reactors, the roots are almost exposed to air or another gas mixture (Vaghari et al. 2017). Researchers have scaled up the production of valuable secondary metabolites and/or hairy roots of medicinally important plants in various types of bioreactors like the conventional airlift, bubble column, stirred tank, airlift balloon, and nutrient mist bioreactors (Dhiman et al. 2018). The design of the reactor also depends on whether the product is intracellular or extracellular (Vaghari et al. 2017). Vinterhalter et al. (2021) evaluated the biomass production and xanthone concentration in *Gentiana dinarica* hairy root cultures in three simple bioreactors, such as temporary immersion systems RITA® (TIS RITA®), bubble column bioreactors (BCB), and Erlenmeyer flasks (EF); the authors reported that TIS RITA® and BCB containing ½-strength MS medium with 4% sucrose produced higher biomass accumulation, and production of the xanthones norswertianin-1-*O*-primeveroside (nor-1-*O*-prim) and norswertianin than those cultivated in

TABLE 3.3 Synthetic Seeds of Medicinal Plants

Plant Species	Propagule	Matrix	Regeneration (%)	References
Bacopa monnieri	Shoot tips	Sodium alginate (2.5%) MS $CaCl_2$ (75 mM)	100	Pramanik et al. 2021
Solanum trilobatum	Nodal segments	Sodium alginate (3%) MS $CaCl_2$ (80 mM)	98.4	Shilpha et al. 2021
Crinum malabaricum	Somatic embryos	Sodium alginate (2%) MS $CaCl_2$ (100 mM)	92.5	Priyadharshini et al. 2020
Hedychium coronarium	Shoot segments	Sodium alginate (3%) MS $CaCl_2$ (100 mM)	90	Behera et al. 2020
Plectranthus amboinicus	Shoot tips Nodal segments	Sodium alginate (3%) MS $CaCl_2$ (100 mM)	63.3	Arumugam et al. 2019
Althaea officinalis	Nodal segment	Sodium alginate (3%) MS $CaCl_2$ (100 mM)	66	Naz et al. 2018
Capparis decidua	Shoot tips, nodal segments	Sodium alginate (3%) MS $CaCl_2$ (100 mM)	61–91.6	Ahlawat et al. 2018
Salix tetrasperma	Nodal segments	Sodium alginate (3%) WPM $CaCl_2$ (75 mM)	71	Khan et al. 2018
Picrorhiza kurrooa	Hairy roots	Sodium alginate (3%) MS $CaCl_2$ (3%)	73	Rawat et al. 2013
Centaurium erythraea	Hairy roots	Sodium alginate (3%) MS and WPM $CaCl_2$ (50 mM)	56–86	Piątczak and Wysokińska 2013
Plumbago indica	Hairy roots	Sodium alginate (2%) MS $CaCl_2$ (100 mM)	80–90	Gangopadhyay et al. 2011

MS: Murashige and Skoog (1962), WPM: Woody Plant Media (McCown and Lloyd 1981)

EF. The optimal culture conditions for the *G. dinarica* hairy root clones were highly aerated TIS RITA® and BCB systems.

3.10 CONCLUSIONS

The hairy root culture as a biotechnological tool to produce secondary metabolites has been studied over recent decades. Several plant metabolites have been used for the treatment of various diseases and conditions for centuries. The analysis of many plant extracts used in traditional medicine has reported biological activities, such as cytotoxicity of cancer cell lines, antibacterial, fungicidal, anti-inflammatory, or antiviral activities. However, secondary metabolites are often accumulated in specific tissues and at particular stages of development, which make the appropriate extracts difficult to obtain. The hairy root culture is an interesting option to produce metabolites that are naturally synthesized and accumulated in medicinal plants and allows the possibility of having a constant source of rapidly growing biomass containing high concentrations of specific metabolites under controlled conditions independent of both plant growth regulators and environmental conditions. Moreover, because of the specific transformation via *A. rizhogenes*, it could be possible to achieve the biosynthesis of new secondary metabolites that are not detected in non-transformed tissues, given that numerous and different transformed hairy root lines can be generated. The yield in the extraction of valuable chemical compounds could be like those of wild-type plants or could exhibit an improved chemical profile, through the application of appropriate molecular techniques such as elicitors. Hairy root culture offers the opportunity to obtain plants regenerated from hairy root tissue, because of specific plant morphological characteristics, such as decreased apical dominance, dwarfing, increased branching, wrinkled leaves, among others; such phenotypic changes could pique the interest of the ornamental plant industry. Unfortunately, hairy root culture has not generated a successful biotechnological process that can be used at an industrial level due to several problems: the required design of an appropriate bioreactor, low yields, and complications in scaling up the processes, among others. However, this is an interesting technique to produce metabolites specifically biosynthesized in plant roots and even plausible in other tissues, especially for medicinal plants. Many plant species require at least 10–15 years to reach the stage at which to accumulate the valuable metabolites, which can be reduced to months through the application of hairy culture. Finally, this technology is very attractive for the production of valuable metabolites for the pharmaceutical, agronomic, and food industries. It is necessary to carry out several studies about the scaling-up of hairy roots culture, regulation of biosynthesis pathways, and the factors affecting the accumulation of bioactive molecules, to produce valuable chemical compounds. Thus, hairy root cultures are expected to be present in commercial applications in the not-too-distant future.

ACKNOWLEDGMENTS

AADVM acknowledges the financial support provided by SIP Instituto Politécnico Nacional Mexico (IPN/SIP 20211305; IPN/SIP 20220810) and Ma. Asunción Bravo Díaz for technical support.

REFERENCES

Ahlawat J., R. Choudhary, A. R. Sehrawat, and S. Kaur. 2018. "Germplasm conservation via encapsulating *in vitro* generated shoot tips and nodal segment of *Capparis decidua* (FORSK.) EDGEW and its regeneration." *Romanian Biotechnological Letters* 24(5): 783–88. https://doi.org/10.25083/rbl/24.5/783.788.

Alagarsamy K., L. F. Shamala, and S. Wei. 2018. "Protocol: High-efficiency *in-planta Agrobacterium*-mediated transgenic hairy root induction of *Camellia sinensis* var. *sinensis*." *Plant Methods* 14(1): 17. https://doi.org/10.1186/s13007-018-0285-8

Alsoufi A. S. M., K. Staśkiewicz, and M. Markowski. 2021. "Alterations in oleanolic acid and sterol content in marigold (*Calendula officinalis*) hairy root cultures in response to stimulation by selected phytohormones." *Acta Physiologiae Plantarum* 43: 44. https://doi.org/10.1007/s11738-021-03212-6.

Arumugam G., U. R. Sinniah, M. K. Swamy, and P. T. Lynch. 2019. "Encapsulation of *in vitro Plectranthus amboinicus* (Lour.) Spreng. shoot apices for propagation and conservation." *3 Biotech* 9(8): 298. https://doi.org/10.1007/s13205-019-1831-4.

Balakrishna D., R. Vinodh, P. Madhu, S. Avinash, P. V. Rajappa, and B. V. Bhat. 2019. "Tissue culture and genetic transformation in *Sorghum bicolor*." In *Breeding Sorghum for Diverse end Uses*, edited by C. Aruna, K. B. R. S. Visarada, B. V. Bhat, and A. T. Vilas, 115–30. Elsevier. https://doi.org/10.1016/B978-0-08-101879-8.00007-3.

Banihashemi O., R.-A. Khavari-Nejad, N. Yassa, and F. Najafi. 2020. "Raise up of scopolamine in hairy roots via *Agrobacterium rhizogenes* ATCC15834 as compared with untransformed roots in *Atropa komarovii*." *Iranian Journal of Pharmaceutical Research* 19(1): 46–56. https://doi.org/10.22037/ijpr.2019.13550.11710.

Bazaldúa C., A. Cardoso-Taketa, G. Trejo-Tapia, B. Camacho-Diaz, J. Arellano, E. Ventura-Zapata, and M. L. Villarreal. 2019. "Improving the production of podophyllotoxin in hairy roots of *Hyptis suaveolens* induced from regenerated plantlets." *PLoS One* 14(9): e0222464. https://doi.org/10.1371/journal.pone.0222464.

Behera S., K. K. Rout, P. C. Panda, and S. K. Naik. 2020. "Production of non-embryogenic synthetic seeds for propagation and germplasm transfer of *Hedychium coronarium* J. Koenig." *Journal of Applied Research on Medicinal and Aromatic Plants* 19: 100271. https://doi.org/10.1016/j.jarmap.2020.100271.

Brijwal L., and S. Tamta. 2015. "*Agrobacterium rhizogenes* mediated hairy root induction in endangered *Berberis aristata* DC." *Springerplus* 4: 443. https://doi.org/10.1186/s40064-015-1222-1.

Chen K., and L. Otten. 2017. "Natural *Agrobacterium* transformants: Recent results and some theoretical considerations." *Frontiers in Plant Science* 8: 1600. https://doi.org/10.3389/fpls.2017.01600.

Chung I.-M., K. Rekha, G. Rajakumar, and M. Thiruvengadam. 2016. "Production of glucosinolates, phenolic compounds and associated gene expression profiles of hairy root cultures in turnip (*Brassica rapa* ssp. *rapa*)." *Biotech* 6(2): 175. https://doi.org/10.1007/s13205-016-0492-9.

Desmet S., E. Dhooghe, E. De Keyser, P. Quataert, T. Eeckhaut, J. Van Huylenbroeck, and D, Geelen. 2020. "Segregation of *rol* genes in two generations of *Sinningia speciosa* engineered through wild type *Rhizobium rhizogenes*." *Frontiers in Plant Science* 11: 859. https://doi.org/10.3389/fpls.2020.00859.

Dessaux Y., and D. Faure. 2018. "Niche construction and exploitation by *Agrobacterium*: How to survive and face competition in soil and plant habitats." In *Agrobacterium Biology*, edited by S. B. Gelvin, 55–86. Cham: Springer. https://doi.org/10.1007/82_2018_83.

Dhiman N., V. Patial, and A. Bhattacharya. 2018. "The current status and future applications of hairy root cultures." In *Biotechnological Approaches for Medicinal and Aromatic Plants*, edited by N. Kumar, 87–155. Singapore: Springer. https://doi.org/10.1007/978-981-13-0535-1_5.

Eungsuwan N., P. Chayjarung, J. Pankam, V. Pilaisangsuree, P. Wongshaya, A. Kongbangkerd, C. Sriphannam and A. Limmongkon. 2021. "Production and antimicrobial activity of *trans*-resveratrol, *trans*-arachidin-1 and *trans*-arachidin-3 from elicited peanut hairy root cultures in shake flasks compared with bioreactors." *Journal of Biotechnology* 326: 28–36. https://doi.org/10.1016/j.jbiotec.2020.12.006.

Gamborg O. L., R. A. Miller, K. Ojima. 1968. "Nutrient requirements of suspension cultures of soybean root cells." *Experimental Cell Research* 50(1): 151–8 https://doi.org/10.1016/0014-4827(68)90403-5.

Gangopadhyay M., S. Dewanjee, D. Chakraborty, and S. Bhattacharya. 2011. "Role of exogenous phytohormones on growth and plumbagin accumulation in *Plumbago indica* hairy roots and conservation of elite root clones via synthetic seeds." *Industrial Crops and Products* 33(2): 445–50. https://doi.org/10.1016/j.indcrop.2010.10.030.

Gantait S., and E. Mukherjee. 2021. "Hairy root culture technology: Applications, constraints and prospect." *Applied Microbiology and Biotechnology* 105: 35–53. https://doi.org/10.1007/s00253-020-11017-9.

Garagounis C., K. Beritza, M.-E. Georgopoulou, P. Sonawane, K. Haralampidis, A. Goossens, A. Aharoni, K. K. Papadopoulou. 2020. "A hairy-root transformation protocol for *Trigonella foenum-graecum* L. as a tool for metabolic engineering and specialised metabolite pathway elucidation." *Plant Physiology and Biochemistry* 154: 451–62. https://doi.org/10.1016/j.plaphy.2020.06.011.

Gelvin S. B. 2017. "Integration of *Agrobacterium* T-DNA into the plant genome." *Annual Review of Genetics* 51: 195–217. https://doi.org/10.1146/annurev-genet-120215-035320.

Guo M., Y. Jingyang, G. Dawei, X. Nan, and Y. Jing. 2019. "*Agrobacterium*-mediated horizontal gene transfer: Mechanism, biotechnological application, potential risk and forestalling strategy." *Biotechnology Advances* 37(1): 259–70. https://doi.org/10.1016/j.biotechadv.2018.12.008.

Gutierrez-Valdes N., S. T. Häkkinen, C. Lemasson, M. Guillet, K.-M. Oksman-Caldentey, A. Ritala, and F. Cardon. 2020. "Hairy root cultures - A versatile tool with multiple applications." *Frontiers in Plant Science* 11: 33. https://doi.org/10.3389/fpls.2020.00033.

Habibi P., M. F. G. De Sa, A. L. L. Da Silva, A. Makhzoum, J. D. L. Costa, I. A. Borghetti, and C. R. Soccol. 2016. "Efficient genetic transformation and regeneration system from hairy root of *Origanum vulgare*." *Physiology and Molecular Biology of Plants* 22(2): 271–77. https://doi.org/10.1007/s12298-016-0354-2.

Halder M., D. Roychowdhury, and S. Jha. 2018. "A critical review on biotechnological interventions for production and yield enhancement of secondary metabolites in hairy root cultures." In *Hairy Roots*, edited by V. Srivastava, S. Mehrotra, and S. Mishra, 21–44. Singapore: Springer. https://doi.org/10.1007/978-981-13-2562-5_2.

Hanafy M. S., M. S. Asker, H. El-Shabrawi, and M. A. Matter. 2018. "*Agrobacterium rhizogenes*-mediated genetic transformation in *Cichorium* spp.: Hairy root production, inulin and total phenolic compounds analysis." *The Journal of Horticultural Science and Biotechnology* 93(6): 605–13. https://doi.org/10.1080/14620316.2017.1420429.

Ho T.-T., J.-D. Lee, M.-S. Ahn, S.-W. Kim, and S.-Y. Park. 2018. "Enhanced production of phenolic compounds in hairy root cultures of *Polygonum multiflorum* and its metabolite discrimination using HPLC and FT-IR methods." *Applied Microbiology and Biotechnology* 102(22): 9563–75. https://doi.org/10.1007/s00253-018-9359-9.

Hu J., S. Liu, Q. Cheng, S. Pu, M. Mao, Y. Mu, F. Dan, J. Yang, and M. Ma. 2020. "Novel method for improving ardicrenin content in hairy roots of *Ardisia crenata* Sims plants." *Journal of Biotechnology* 311: 12–8. https://doi.org/10.1016/j.jbiotec.2020.02.009.

Huang P., L. Xia, W. Liu, R. Jiang, X. Liu, Q. Tang, M. Xu, L. Yu, Z. Tang, and J. Zeng. 2018. "Hairy root induction and benzylisoquinoline alkaloid production in *Macleaya cordata*." *Scientific Reports* 8(1): 11986. https://doi.org/10.1038/s41598-018-30560-0.

Jamwal K., S. Bhattacharya, and S. Puri. 2018. "Plant growth regulator mediated consequences of secondary metabolites in medicinal plants." *Journal of Applied Research on Medicinal and Aromatic Plants* 9: 26–38. https://doi.org/10.1016/j.jarmap.2017.12.003.

Kanchanamala R. W. M. K., and P. C. G. Bandaranayake. 2019. "An efficient and rapid *Rhizobium rhizogenes* root transformation protocol for *Lemna minor*." *Plant Biotechnology Reports* 13: 625–33. https://doi.org/10.1007/s11816-019-00558-9.

Khan M. I., N. Ahmad, M. Anis, A. A. Alatar, and M. Faisal. 2018. "*In vitro* conservation strategies for the Indian willow (*Salix tetrasperma* Roxb.), a vulnerable tree species via propagation through synthetic seeds." *Biocatalysis and Agricultural Biotechnology* 16: 17–21. https://doi.org/10.1016/j.bcab.2018.07.002.

Khazaei A., B. Bahramnejad, A.-A. Mozafari, D. Dastan, and S. Mohammadi. 2019. "Hairy root induction and farnesiferol b production of endemic medicinal plant *Ferula pseudalliacea*." *3 Biotech* 9(11): 407. https://doi.org/10.1007/s13205-019-1935-x.

Kocak M., B. Sevindik, T. Izgu, M. Tutuncu, and Y. Y. Mendi. 2019. "Synthetic seed production of flower bulbs." In *Synthetic Seeds*, edited by M. Faisal and A. A. Alatar, 283–99. Cham: Springer International Publishing. https://doi.org/10.1007/978-3-030-24631-0_12.

Kochan E., A. Nowak, M. Zakłos-Szyda, D. Szczuka, G. Szymanska, and I. Motyl. 2020. "*Panax quinquefolium* L. ginsenosides from hairy root cultures and their clones exert cytotoxic, genotoxic and pro-apoptotic activity towards human colon adenocarcinoma cell line Caco-2." *Molecules* 25: 2262. https://doi.org/10.3390/molecules25092262.

Kumar A. M., S. S. S. Pammi, M. S. Sukanya, and A. Giri. 2018. "Enhanced production of pharmaceutically important isoflavones from hairy root rhizoclones of *Trifolium pratense* L." *In Vitro Cellular and Developmental Biology - Plant* 54(1): 94–103. https://doi.org/10.1007/s11627-017-9873-y.

Lima J. E., V. A. Benedito, A. Figueira, and L. E. Peres. 2009. "Callus, shoot and hairy root formation *in vitro* as affected by the sensitivity to auxin and ethylene in tomato mutants." *Plant Cell Reports* 28(8): 1169–77. https://doi.org/10.1007/s00299-009-0718-y.

Martínez M. I., G. Barba-Espín, B. T. Favero, and H. Lütken. 2020. "*Rhizobium rhizogenes*-mediated transformation of *Rhodiola rosea* leaf explants." *Bragantia* 79(2): 213–23. https://doi.org/10.1590/1678-4499.20190428.

Matveeva T., and L. Otten. 2021. "Opine biosynthesis in naturally transgenic plants: Genes and products." *Phytochemistry* 189: 112813. https://doi.org/10.1016/j.phytochem.2021.112813.

McCown B. H., and G. Lloyd. 1981. "Woody plant medium (WPM) a mineral nutrient formulation for microculture for woody plant species." *Horticultural Science* 16: 453.

Mehrotra S., S. Mishra, and V. Srivastava. 2018. "Bioreactor technology for hairy roots cultivation." In *Bioprocessing of Plant in Vitro Systems*, edited by A. Pavlov and T. Bley, 483–506. Cham: Springer. https://doi.org/10.1007/978-3-319-54600-1_10.

Mikhaylova E. V., A. Artyukhin, K. Musin, M. Panfilova, G. Gumerova, and B. Kuluev. 2021. "The first report on the induction of hairy roots in *Trapa natans*, a unique aquatic plant with photosynthesizing roots." *Plant Cell, Tissue and Organ Culture* 144: 485–90. https://doi.org/10.1007/s11240-020-01963-7.

Moehninsi, D. A. Navarre 2018. "Optimization of hairy root induction in *Solanum tuberosum. American Journal of Potato Research* 95: 650–8 https://doi.org/10.1007/s12230-018-9671-z

Murashige T. 1977. "Plant cell and organ cultures as horticultural practices." *Acta Horticulturae* 78: 17–30. https://doi.org/10.17660/ActaHortic.1977.78.1

Murashige T., and F. Skoog. 1962. "A revised medium for rapid growth and bio assays with tobacco tissue cultures." *Physiologia Plantarum* 15: 473–97. https://doi.org/10.1111/j.1399-3054.1962.tb08052.x.

Muthusamy, B., and G. Shanmugam. 2020. "Analysis of flavonoid content, antioxidant, antimicrobial and antibiofilm activity of *in vitro* hairy root extract of radish (*Raphanus sativus* L.)." *Plant Cell, Tissue and Organ Culture* 140: 619–33. https://doi.org/10.1007/s11240-019-01757-6.

Nakasha J. J., U. R. Sinniah, N. A. Shaharuddin, S. A. Hassan, S. Subramaniam, and M. K. Swamy. 2017. "Establishment of an efficient *in vitro* regeneration and *Agrobacterium rhizogenes*-mediated genetic transformation protocol for safed musli (*Chlorophytum borivilianum* Santapau & R.R.Fern.)." *In Vitro Cellular and Developmental Biology: Plant* 53(6): 571–8. https://doi.org/10.1007/s11627-017-9831-8.

Navarre, D. A. 2018. "Optimization of hairy root induction in *Solanum tuberosum*." *American Journal of Potato Research* 95: 650–8. https://doi.org/10.1007/s12230-018-9671-z.

Naz R., M. Anis, A. A. Alatar, A. Ahmad, and A. Naaz. 2018. "Nutrient alginate encapsulation of nodal segments of *Althaea officinalis* L., for short-term conservation and germplasm exchange." *Plant Biosystems: An International Journal Dealing With All Aspects of Plant Biology* 152(6): 1256–62. https://doi.org/10.1080/11263504.2018.1436610.

Neumann M., S. Prahl, L. Caputi, L. Hill, B. Kular, A. Walter, E. P. Patallo, D. Milbredt, A. Aires, M. Schöpe, S. O'Connor, K.-H. van Pée, and J. Ludwig-Müller. 2020. "Hairy root transformation of *Brassica rapa* with bacterial halogenase genes and regeneration to adult plants to modify production of indolic compounds." *Phytochemistry* 175: 112371. https://doi.org/10.1016/j.phytochem.2020.112371.

Nourozi E., B. Hosseini, R. Maleki, and B. A. Mandoulakani. 2019. "Pharmaceutical important phenolic compounds overproduction and gene expression analysis in *Dracocephalum kotschyi* hairy roots elicited by SiO_2 nanoparticles." *Industrial Crops and Products* 133: 435–46. https://doi.org/10.1016/j.indcrop.2019.03.053.

Otten L. 2018. "The *Agrobacterium* phenotypic plasticity (*plast*) genes." In *Agrobacterium Biology*, edited by S. B. Gelvin, 375–419. Cham: Springer. https://doi.org/10.1007/82_2018_93.

Otten L. 2021. "TDNA regions from 350 *Agrobacterium* genomes: Maps and phylogeny." *Plant Molecular Biology* 106: 239–58. https://doi.org/10.1007/s11103-021-01140-0.

Padilla R., V. Gaillard, T. N. Le, F. Bellvert, D. Chapulliot, X.. Nesme, Y. Dessaux, L. Vial, C. Lavire, and I. Kerzaon. 2021. "Development and validation of a UHPLC-ESI-QTOF mass spectrometry method to analyze opines, plant biomarkers of crown gall or hairy root diseases." *Journal of Chromatography B* 1162: 122458. https://doi.org/10.1016/j.jchromb.2020.122458.

Pedreño M. A., and L. Almagro. 2020. "Carrot hairy roots: Factories for secondary metabolite production." *Journal of Experimental Botany* 71(22): 6861–4. https://doi.org/10.1093/jxb/eraa435.

Piątczak E., and H. Wysokińska. 2013. "Encapsulation of *Centaurium erythraea* Rafn - An efficient method for regeneration of transgenic plants." *Acta Biologica Cracoviensia Series Botanica* 55(2). https://doi.org/10.2478/abcsb-2013-0022.

Pramanik B., S. Sarkar, S. Bhattacharyya, and S. Gantait. 2021. "Meta-Topolin-induced enhanced biomass production via direct and indirect regeneration, synthetic seed production, and genetic fidelity assessment of *Bacopa monnieri* (L.) Pennell, a memory-booster plant." *Acta Physiologiae Plantarum* 43: 107. https://doi.org/10.1007/s11738-021-03279-1.

Priyadharshini, S., M. Manokari, and M. S. Shekhawat. 2020. "*In vitro* conservation strategies for the critically endangered Malabar River lily (*Crinum malabaricum* Lekhak & Yadav) using somatic embryogenesis and synthetic seed production." *South African Journal of Botany* 135: 172–80. https://doi.org/10.1016/j.sajb.2020.08.030.

Qahtan A. A., E. M. Abdel-Salam, A. A. Alatar, Q.-C. Wang, and M. Faisal. 2019. "An introduction to synthetic seeds: Production, techniques, and applications." In *Synthetic Seeds*, edited by M. Faisal and A. A. Alatar, 1–20. Cham: Springer International Publishing. https://doi.org/10.1007/978-3-030-24631-0_1.

Rana M. M., Z.-X. Han, D.-P. Song, G.-F. Liu, D.-X. Li, X.-C. Wan, A. Karthikeyan, and S. Wei. 2016. "Effect of medium supplements on *Agrobacterium rhizogenes* mediated hairy root induction from the callus tissues of *Camellia sinensis* var. *sinensis*." *International Journal of Molecular Sciences* 17(7): 1132. https://doi.org/10.3390/ijms17071132.

Rawat J. M., A. Bhandari, M. Raturi, and B. Rawat. 2019. "*Agrobacterium rhizogenes* mediated hairy root cultures: A promising approach for production of useful metabolites." In *New and Future Developments in Microbial Biotechnology and Bioengineering*, edited by V. K. Gupta and A. Pandey, 103–18. Elsevier. https://doi.org/10.1016/B978-0-444-63504-4.00008-6.

Rawat J. M., B. Rawat, and S. Mehrotra. 2013. "Plant regeneration, genetic fidelity, and active ingredient content of encapsulated hairy roots of *Picrorhiza kurrooa* Royle Ex Benth." *Biotechnology Letters* 35(6): 961–8. https://doi.org/10.1007/s10529-013-1152-3.

Rihan H., F. Kareem, M. El-Mahrouk, and M. Fuller. 2017. "Artificial seeds (principle, aspects and applications)." *Agronomy* 7(4): 71. https://doi.org/10.3390/agronomy7040071.

Ruiz-Ramírez L. A., G. del C. Godoy-Hernández, P. E. Álvarez-Gutiérrez, V. M. Ruíz-Valdiviezo, M. C. Lujan-Hidalgo, E. Avilés-Berzunza, and F. A. Gutiérrez-Miceli. 2018. "Protocol for bonediol production in *Bonellia macrocarpa* hairy root culture." *Plant Cell, Tissue and Organ Culture* 134(1): 177–81. https://doi.org/10.1007/s11240-018-1403-z.

Saeger J. D., J. Park, H. S. Chung, J.-P. Hernalsteens, M. Van Lijsebettens, D. Inzé, M. Van Montagu, and S. Depuydt. 2020. "*Agrobacterium* strains and strain improvement: Present and outlook." *Biotechnology Advances* 53: 107677. https://doi.org/10.1016/j.biotechadv.2020.107677.

Sahai P., and V. B. Sinha. 2021. "Development of hairy root culture in *Taxus baccata* sub sp *wallichiana* as an alternative for increased taxol production." *Materials Today: Proceedings*. https://doi.org/10.1016/j.matpr.2021.03.407.

Sahayarayan J. J., R. Udayakumar, M. Arun, A.. Ganapathi, M. S. Alwahibi, N. S. Aldosari, and A. M. A. Morgan. 2020. "Effect of different *Agrobacterium rhizogenes* strains for *in-vitro* hairy root induction, total phenolic, flavonoids contents, antibacterial and antioxidant activity of (*Cucumis anguria* L.)." *Saudi Journal of Biological Sciences* 27(11): 2972–9. https://doi.org/10.1016/j.sjbs.2020.08.050.

Sarkar J., A. Misra, and N. Banerjee. 2020. "Genetic transfection, hairy root induction and solasodine accumulation in elicited hairy root clone of *Solanum erianthum* D. Don." *Journal of Biotechnology* 323: 238–45. https://doi.org/10.1016/j.jbiotec.2020.09.002.

Sarkar S., I. Ghosh, D. Roychowdhury, and S. Jha. 2018. "The effects of *rol* genes of *Agrobacterium rhizogenes* on morphogenesis and secondary metabolite accumulation in medicinal plants." In *Biotechnological Approaches for Medicinal and Aromatic Plants*, edited by N. Kumar, 27–51. Singapore: Springer. https://doi.org/10.1007/978-981-13-0535-1_2.

Savić J., R. Nikolić, N. Banjac, S. Zdravković-Korać, S. Stupar, A. Cingel, T. Ćosić, M. Raspor, A. Smigocki, and S. Ninković. 2019. "Beneficial implications of sugar beet proteinase inhibitor BvSTI on plant architecture and salt stress tolerance in *Lotus corniculatus* L." *Journal of Plant Physiology* 243: 153055. https://doi.org/10.1016/j.jplph.2019.153055.

Shilpha J., S. Pandian, M. J. V. Largia, S. I. Sohn, and M. Ramesh. 2021. "Short-term storage of *Solanum trilobatum* L. synthetic seeds and evaluation of genetic homogeneity using SCoT markers." *Plant Biotechnology Reports* 15: 651–61. https://doi.org/10.1007/s11816-021-00709-x.

Singh R. S., T. Chattopadhyay, D. Thakur, N. Kumar, T. Kumar, and P. K. Singh. 2018. "Hairy root culture for *in vitro* production of secondary metabolites: A promising biotechnological approach." In *Biotechnological Approaches for Medicinal and Aromatic Plants*, edited by N. Kumar, 235–50. Singapore: Springer. https://doi.org/10.1007/978-981-13-0535-1_10.

Singh R. S., V. K. Jha, T. Chattopadhyay, U. Kumar, D. P. Fulzele, and P. K. Singh. 2020. "First report of *Agrobacterium rhizogenes*-induced hairy root formation in *Selaginella bryopteris*: A Pteridophyte recalcitrant to genetic transformation." *Brazilian Archives of Biology and Technology* 63: e20180679. http://dx.doi.org/10.1590/1678-4324-2020180679.

Skała E., A. Kicel, M. A. Olszewska, A. K. Kiss, and H. Wysokińska. 2015. "Establishment of hairy root cultures of *Rhaponticum carthamoides* (Willd.) Iljin for the production of biomass and caffeic acid derivatives." *Biomed Research International* 2015: 181098 https://doi.org/10.1155/2015/181098.

Srikantan C., and S. Srivastava. 2018. "Bioreactor design and analysis for large-scale plant cell and hairy root cultivation." In *Hairy Roots*, edited by V. Srivastava, S. Mehrotra, and S. Mishra, 147–82. Singapore: Springer. https://doi.org/10.1007/978-981-13-2562-5_7.

Srivastava V., S. Mehrotra, and P. K. Verma. 2017. "Biotechnological interventions for production of therapeutic secondary metabolites using hairy root cultures of medicinal plants." In *Current Developments in Biotechnology and Bioengineering*, edited by S. K. Dubey, A. Pandey, and R. S. Sangwan, 259–82. Elsevier. https://doi.org/10.1016/B978-0-444-63661-4.00012-8.

Stavnstrup S. S., J. P. Molina, H. Lütken, R. Müller, and J. N. Hegelund. 2020. "Ancient horizontal gene transfer from *Rhizobium* rhizogenes to European genera of the Figwort family (Scrophulariaceae)." *Euphytica* 216: 186. https://doi.org/10.1007/s10681-020-02722-7.

Sun L., Y. Ge, J. A. Sparks, Z. T. Robinson, X. Cheng, J. Wen, and E. B. Blancaflor. 2019. "TDNAscan: A software to identify complete and truncated T-DNA insertions." *Frontiers in Genetics* 10: 685. https://doi.org/10.3389/fgene.2019.00685.

Tanaka N., M. Yamakawa, and I. Yamashita. 1998. "Characterization of transcription of genes involved in hairy root induction on pRi1724 core-T-DNA in two *Ajuga reptans* hairy root lines." *Plant Science* 137(1): 95–105. https://doi.org/10.1016/s0168-9452(98)00123-x.

Tatsumi K., T. Ichino, N. Onishi, K. Shimomura, and K. Yazaki. 2020. "Highly efficient method of *Lithospermum erythrorhizon* transformation using domestic *Rhizobium rhizogenes* strain A13." *Plant Biotechnology* 37: 39–46. https://doi.org/10.5511/plantbiotechnology.19.1212a.

Tavassoli P., and A. S. Afshar. 2018. "Influence of different *Agrobacterium rhizogenes* strains on hairy root induction and analysis of phenolic and flavonoid compounds in marshmallow (*Althaea officinalis* L.)." *3 Biotech* 8(8): 351. https://doi.org/10.1007/s13205-018-1375-z.

Thakur M., S. Bhattacharya, P. K. Khosla, and S. Puri. 2019. "Improving production of plant secondary metabolites through biotic and abiotic elicitation." *Journal of Applied Research on Medicinal and Aromatic Plants* 12: 1–12. https://doi.org/10.1016/j.jarmap.2018.11.004.

Thilip, C., C. S. Raju, K. Varutharaju, A. Aslam, and A. Shajahan. 2015. "Improved *Agrobacterium rhizogenes*-mediated hairy root culture system of *Withania somnifera* (L.) Dunal using sonication and heat treatment." *3 Biotech* 5: 949–56. https://doi.org/10.1007/s13205-015-0297-2.

Thiruvengadam M., K. Rekha, and I.-M. Chung. 2016. "Induction of hairy roots by *Agrobacterium rhizogenes*-mediated transformation of spine gourd (*Momordica dioica* Roxb. ex. willd) for the assessment of phenolic compounds and biological activities." *Scientia Horticulturae* 198: 132–41. https://doi.org/10.1016/j.scienta.2015.11.035.

Vaghari H., H. Jafarizadeh-Malmiri, N. Anarjan, and A. Berenjian. 2017. "Hairy root culture: A biotechnological approach to produce valuable metabolites." In *Agriculturally Important Microbes for Sustainable Agriculture*, edited by V. Meena, P. Mishra, J. Bisht, and A. Pattanayak. Singapore: Springer. https://doi.org/10.1007/978-981-10-5589-8_7.

Vamenani R., A. Pakdin-Parizi, M. Mortazavi, and Z Gholami. 2020. "Establishment of hairy root cultures by *Agrobacterium rhizogenes* mediated transformation of *Trachyspermum ammi* L. for the efficient production of thymol." *Biotechnology and Applied Biochemistry* 67(3): 389–95. https://doi.org/10.1002/bab.1880.

Vergara-Martínez, V. M., S. E. Estrada-Soto, S. Valencia-Díaz, K. Garcia-Sosa, L. M. Peña-Rodríguez, J. de J. Arellano-García, and I. Perea-Arango. 2021. "Methyl jasmonate enhances ursolic, oleanolic and rosmarinic acid production and sucrose induced biomass accumulation, in hairy roots of *Lepechinia caulescens*." *PeerJ* 9: e11279. http://doi.org/10.7717/peerj.11279.

Vinterhalter B., J. Savić, S. Zdravković-Korać, N. Banjac, D. Vinterhalter, and D. Krstić-Milošević. 2019. "*Agrobacterium rhizogenes*-mediated transformation of *Gentiana utriculosa* L. and xanthones decussatin-1-*O*-primeveroside and decussatin accumulation in hairy roots and somatic embryo-derived transgenic plants." *Industrial Crops and Products* 130: 216–29. https://doi.org/10.1016/j.indcrop.2018.12.066.

Vinterhalter B., N. Banjac, D. Vinterhalter, and D. Krstić-Milošević. 2021. "Xanthones production in *Gentiana dinarica* Beck hairy root cultures grown in simple bioreactors." *Plants* 10: 1610. https://doi.org/10.3390/plants10081610.

Wojciechowska M., A. Owczarek, A. K. Kiss, R. Grąbkowska, M. A. Olszewska, and I. Grzegorczyk-Karolak. 2020. "Establishment of hairy root cultures of *Salvia bulleyana* Diels for production of polyphenolic compounds." *Journal of Biotechnology* 318: 10–9. https://doi.org/10.1016/j.jbiotec.2020.05.002.

Wons E., D. Koscielniak, M. Szadkowska, and M. Sektas. 2018. "Evaluation of GFP reporter utility for analysis of transcriptional slippage during gene expression." *Microbial Cell Factories* 17: 150. https://doi.org/10.1186/s12934-018-0999-3.

Zhang B., M. Chen, S. Pu, L. Chen, X. Zhang, J. Zhang, and C. Zhu. 2020. "Identification of secondary metabolites in *Tripterygium wilfordii* hairy roots and culture optimization for enhancing wilforgine and wilforine production." *Industrial Crops and Products* 148: 112276. https://doi.org/10.1016/j.indcrop.2020.112276.

Zhao S., and H. Tang. 2020. "Enhanced production of valtrate in hairy root cultures of *Valeriana jatamansi* Jones by methyl jasmonate, jasmonic acid and salicylic acid elicitors." *Notulae Botanicae Horti Agrobotanici Cluj-napoca* 48(2): 839–48. https://doi.org/10.15835/nbha48211891.

Zhong C., M. Nambiar-Veetil, D. Bogusz, and C. Franche. 2018. "Hairy roots as a tool for the functional analysis of plant genes." In *Hairy Roots*, edited by V. Srivastava, S. Mehrotra, and S. Mishra, 275–92. Singapore: Springer. https://doi.org/10.1007/978-981-13-2562-5_12.

Zolfaghari F., S. Rashidi-Monfared, A. Moieni, D. Abedini, and A. Ebrahimi. 2020. "Improving diosgenin production and its biosynthesis in *Trigonella foenum-graecum* L. hairy root cultures." *Industrial Crops and Products* 145: 112075. https://doi.org/10.1016/j.indcrop.2019.112075.

CHAPTER 4

Differentiated Plant Cell Suspension Cultures: The New Promise for Secondary Metabolite Production

Daniel Arturo Zavala-Ortíz,
Javier Gómez-Rodríguez, and
María Guadalupe Aguilar-Uscanga

CONTENTS

DOI: 10.1201/9781003166535-4

4.1 INTRODUCTION

It has been a long time since humans realized that some plants could provide relief of illness or act as a poison for enemies. There are molecules within plants which exhibit such desired bioactivities. Usually, these bioactive molecules comprise less than 1% of the total plant carbon and are therefore called secondary metabolites in contrast to the most abundant primary or principal metabolites. Even today, there is not a universal definition of secondary metabolite, so that the concept within plant cell sciences varies among authors. For instance, secondary metabolites have been defined (Verpoorte and Alfermann 2000) as molecules that are not necessary for plant survival but which play a critical role in plant interactions with their environment. Moreover, such molecules are usually restricted to particular organs, tissues, or cells within the plants. Accordingly, there is a vast chemical diversity of secondary metabolites, with biological functions according to the physiological needs of the plants.

Modern identification of secondary metabolites with therapeutic effects comprises a long process requiring several years of intensive research. Perfect examples are the identification of anticancer molecules, such as paclitaxel in *Taxus* plants or vincristine and vinblastine in *Catharanthus* herbs. For instance, the path to the discovery of paclitaxel began in the 1950s in the United States (Suffness 1995). In 1955, the US Congress commissioned the National Cancer Institute (NCI) to organize a program for the collection and identification of therapeutic agents from plants through an agreement with the US Department of Agriculture (USDA). This program resulted in the analysis of *Taxus brevifolia* (Pacific yew) samples, which showed cytotoxic activity toward the KB cell line, from an epidermal mouth carcinoma. *In vivo* antineoplastic activity was also found against a class of leukemia cells, Walker 526 carcinosarcoma, among others. Later, a pseudo-alkaloid was identified as the active principle of the crude extract of the Pacific yew bark and called paclitaxel (Shao et al. 2021). The development of paclitaxel as a medicine required cooperation between the NCI and Bristol-Myers Squibb (BMS) in the 1980s. Finally, BMS received approval from the US Food and Drug Administration (USFDA) for paclitaxel to be used in the treatment of ovarian cancer in 1992, and later for treatment of other types of cancer. *Catharanthus roseus* plants, used in traditional medicines, acquired relevance during screening for molecules with pharmaceutical potential, and the development of vincristine and vinblastine as bioactive molecules against cancer started in the 1950s (Noble 1990).

The former examples depict the hard work necessary for a metabolite to be approved by medicine regulatory agencies. Nowadays, there is a large and increasing list of natural plant secondary metabolites with pharmacological potential. Moreover, the design of new semi-synthetic metabolites is an exciting domain to produce molecules with novel or enhanced pharmacological properties; however, there remain limitations of supply as concentrations in plants are low. As a consequence, new supply alternatives have emerged. In this chapter, the current state of such alternatives is described, with a focus on plant cell suspension cultures as a strategy with promising potential for industrial scale-up.

Once the activity of a metabolite is shown to be promising for therapeutic purposes, it must be accordingly supplied for the clinical trials. Provided that the molecule receives the approval from medicine regulatory agencies, large-scale supply must be guaranteed for commercial issues. However, this is a challenging task for secondary metabolites. For instance, the concentration *in planta* of paclitaxel can be as low as 0.01% of the bark dry weight (Shao et al. 2021) in *Taxus* trees. Moreover, *Taxus* trees are slow-growing and require around 100 years for the bark to accumulate the paclitaxel within. If the bark is harvested from 100-year-old trees, the viability of the harvested trees is compromised. Therefore, novel alternatives must be considered to fulfill the commercial demands. Currently, there are different technological approaches for the large-scale production of secondary metabolites: extraction from plant material, semi-synthesis, chemical synthesis, cell culture-derived processes, and microbial processes (Flores-Bustamante et al. 2010). The use of a particular approach depends on technical and economic aspects; in this chapter, both issues are addressed for each technological approach, using case analysis. Then, the focus is placed on plant cell culture to produce secondary metabolites. Novel approaches are analyzed and discussed to surpass the current limitations of such technology for the industrial production of plant secondary metabolites.

4.2 EXTRACTION FROM PLANT MATERIAL: CHALLENGING PURIFICATIONS AND LIMITED PROFITS

Extraction of metabolites from plant material is generally the first step for large-scale production. The complex and expensive purification process and the low concentration of metabolites are the major challenges for industrial success. For instance, in March 2019, TEVA Pharmaceuticals informed the US FDA (US Food and Drug Administration) of the decision to discontinue the production of generic vincristine, a pediatric oncological without a proper substitute. TEVA claimed that its decision to leave the market was motivated by low profits in generic vincristine (Palmer 2019a) and the need of the company to increase margins (Palmer 2019b). Later, in 2019 and 2020, there were vincristine shortages in North America (Barrett 2019). This fact pointed out the need for novel biotechnological processes for the production of such important plant antileukemics. For instance, paclitaxel is currently produced by biotechnological platforms to replace production from *Taxus* foliage (Ganem and Franke 2007); it seems likely that new processes will emerge to substitute production by extraction from plant tissues for many other plant molecules.

4.3 SEMI-SYNTHESIS: CHEMICAL AND BIOLOGICAL APPROACHES

Semi-synthesis is the chemical synthesis that uses molecules (precursors or structurally similar molecules) isolated from natural sources (particularly foliage, cell cultures, among others) as the starting materials to produce the molecule of interest. Currently, semi-synthesis is the preferred option for extremely complex molecules such as paclitaxel or vincristine. For instance, the major production of paclitaxel is performed by semi-synthesis from primary taxanes such as baccatin III, 10-deacetylbaccatine III, among others, isolated from yew foliage (Ganem and Franke 2007). On the other hand, due to the molecular complexity and scarcity of vincristine and vinblastine within plants ($\approx$0.00015% DW), large-scale production is mainly carried out from the coupling of the main precursors (catharanthine and vindoline) (Alam et al. 2017). Dependence on climatic conditions could compromise the production of Active Pharmaceutical Ingredients (APIs) production via semi-synthesis, as has been the case for vincristine and vinblastine from *Catharantus* plants (Alam et al. 2017). Another outstanding example of semi-synthesis for large-scale production was the case of Syntex, located in Mexico during the 1940s–1970s, which produced steroid hormones (i.e., progesterone for pregnancy control therapies, among others) from plant precursors; this technology became obsolete as new production techniques emerged (Di Renzo et al. 2020).

4.3.1 Biological Synthesis: Microorganisms

Nowadays, the use of microorganisms is an attractive alternative strategy to produce plant molecules. Several reports demonstrate that some fungi or bacteria are capable of producing medicinal plant secondary metabolites or key intermediates (Kaur et al. 2021). Processes based on microorganisms have several advantages over plant cell cultures, for instance, shorter doubling times and greater tolerance of hydrodynamic stress, among others; therefore, they are more easily cultured in large-scale processes. For vincristine and vinblastine production, it has been reported that endophytic fungi isolated from *Catharanthus roseus* can produce vincristine and vinblastine with titers of vincristine (67 μg/L) and vinblastine (70 μg/L) in the culture broth (Palem et al. 2015). *Eutypella* fungus has been also reported as a vincristine producer, with titers of around 53 μg/L (Kuriakose et al. 2016). Though there are promising results of the production of plant metabolites by endophytes, this technology is not practical for large-scale production. Indeed, there are examples where more complex cell cultures are preferred. For instance, paclitaxel is currently produced by plant cell culture technology in Germany and South Korea, and, as far as can be ascertained, there is not a prospect of an endophyte-based process soon (Blanco Carcache et al. 2021).

Moreover, for paclitaxel production by endophytes, some authors (e.g., McElroy and Jennewein 2018) have claimed that it is unclear whether some microorganisms reported as taxane producers can actually produce taxanes. Indeed, they believe that the detected compounds were more likely artifacts of laboratory culture methods, or compounds misidentified by the analytical methods deployed in the original studies.

4.4 PLANT SUSPENSION CULTURES: CLASSIC AND INNOVATIVE PERSPECTIVES

Some molecules derived from plants are essential for the treatment of several human diseases. Such molecules are often molecularly complex and total chemical synthesis is not an affordable option due to low yields. On the other hand, as biosynthesis is extremely complex, requiring translocation of intermediates between organelles, cells, and tissues, bioavailability in plants is often low. Over the past decades, several works have focused on the enhancement of bioavailability using cell culture technology, particularly suspension cultures. In the remaining parts of this chapter, the basics of suspension culture as well as classic production approaches are summarized with a focus on the challenges limiting the expansion of such processes. Briefly, the production of secondary metabolites *in vivo* is restricted to specialized organs, tissues, or cells, which are likely absent in suspension cultures. Thus, a new perspective is introduced seeking to overcome the "restricted" secondary metabolism in suspension cultures, while also providing fundamental concepts to properly address innovative production processes.

4.4.1 Basics of Suspension Culture Processes

In brief, plant cell suspension culture is the culture of "single" plant cells in liquid medium. However, it is only a euphemism since processes based on single cells within the culture medium are scarcely achieved. Indeed, plant cell suspension cultures are suspensions of aggregates composed of several cells attached to one another. Although this proliferation nature imposes particular challenges (aggregate sedimentation, large gradients, among others), this way of culture permits mimicking of microorganism processes to some extent. Therefore, mechanical agitation under controlled conditions in stainless steel bioreactors could be used on a large scale. The first step for the establishment of large-scale suspension cultures is the generation of a plant cell line as callus (Figure 4.1A). Propagation of callus is carried out by several rounds of subcultures (Figure 4.1B). Then, a portion of actively growing callus is transferred into liquid culture medium under agitation for the induction of callus disaggregation into small clusters, preferably, single cells. Then, only cells in small aggregates or single cells are suspended in the culture medium (Figure 4.1C). This process can be repeated several times through subculture until scaling up in bioreactors (Figure 4.1D).

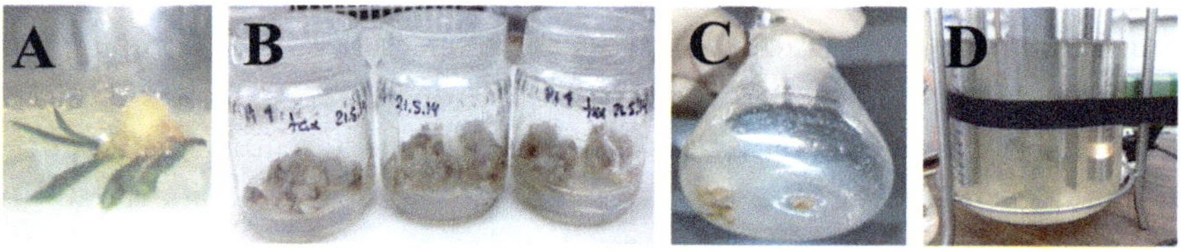

FIGURE 4.1 Different stages of *in vitro* cultures of *Taxus globosa*. Callus induction from stem chunk (A); well-established callus cultures (B); initiation of suspension culture in flask (C); perfusion culture in 3 L bioreactor coupled to a Near-Infrared probe (light-emitting tube) (D).

4.5 HOW TO SUCCEED IN INDUSTRY: EXAMPLES AND LIMITING CHALLENGES OF *IN VITRO* CULTURES

Establishment of industrial processes based on plant cell suspension cultures requires the convergence of economic, technical, and regulatory issues. Consequently, the sale price during the life cycle of the product must sufficiently surpass the initial R&D investment, the production costs, and the medicines approval process, among others, so that investors earn some profits. For large companies in the drug industry, average profits between 15% and 20% are common (US Government Accountability Office (GAO) 2017). In contrast, profits from biopharmaceutical companies are usually higher (Manning and Karki 2019). As a result, the production process must be as competitive as possible. Only high-value molecules in terms of price and demand can yield the proper economic balance for their production by suspension cultures. Generally, such molecules must be scarce in nature, unaffordable by chemical synthesis, and yet prohibitively expensive for production by genetically modified microorganisms. Therefore, an optimized suspension culture process must be developed. However, there can be the case that even a high-yielding cell suspension culture process is insufficient to achieve long-term establishment even if production costs are lower than other production alternatives. For instance, the Mitsui company developed a high-yielding *Lithospermum erythrorhizon* cell suspension culture process to produce shikonin with productions costs (~ 4000 USD kg^{-1}) slightly below the cost of the molecule extracted from the intact plant (Neumann et al. 2020). Then, suspension culture-derived shikonin was launched to market. However, it was soon withdrawn as the low profits could not amortize the costs associated with suspension culture process development (Neumann et al. 2020).

The perfect example of success is the production of paclitaxel using suspension cultures of *Taxus chinensis* in large-scale bioreactors by Phyton Biotech Inc. and Samyang Biopharm in Germany and South Korea, respectively (McElroy and Jennewein 2018). Moreover, there is also a new plant suspension culture process to be launched soon to produce 10-deacetylbaccatin III by Hokkaido Mitsui Chemicals. Currently, there have only been a few successful industrial processes based on plant suspension cultures to produce plant secondary metabolites used in pharmacy, cosmetics, or food industries. For instance, with respect to pharmaceutical products (Lange 2018; Wilson and Roberts 2012), the Mitsui Chemicals Inc portfolio has included the use of plant suspension cultures to produce arbutin from *C. roseus*, berberines from *Coptis japonica* and *Thalictrum minus* (Fujita 2007), geraniol from *Geraminea* spp., and shikonin from *Lithospermum erythrorhizon*; the Nippon Oil portfolio comprises anthocyanins from *Euphorbia milii* and *Aralia cordata*, betacyanins from *Beta vulgaris*, and podophyllotoxin from *Podophyllium* spp.; ginseng is also produced by Nitto Denko Corporation and Unhwa Biotech Corp., using *Panax ginseng*, with rosmarinic acid produced from *Coleus blumei* by Nattermann & Cie, and scopolamine from *Duboisia* spp. by Sumito Chemical Co., Ltd. Perhaps the most extended use of plant suspension cultures has been in the foods and cosmetics industries, where more than 22 processes have been reported (Eibl et al. 2018).

4.6 CLASSIC APPROACH: SECONDARY METABOLISM

In this chapter, an analysis of secondary metabolite definitions is needed. However, caution must be paid since different perspectives are preferred within different disciplines. For instance, in *in vivo* studies, definitions related to ecology and the restricted occurrence in specialized tissues or cells (Verpoorte and Alfermann 2000) are preferred. However, as the term "ecology" in *in vitro* cultures is considered less important, particularly in axenic cultures, secondary metabolites are defined simply as the results of secondary metabolism. Interestingly, secondary metabolism is usually defined as involving the metabolites that are produced by biochemical pathways not directly involved in the growth, development, or reproduction of the cells (Carvalho et al. 2019). Indeed, the latter definition not only disparages the term ecology, but also avoids the implications of the *restricted occurrence*. Restricted occurrence can imply time, location, and nature issues, such as stress events, accumulation in particular organelles, such as vacuoles, or restriction to some differentiated tissue or cells. In *in vitro* techniques of axenic cultures considered to be likely homogeneous, a restricted occurrence is therefore commonly linked only to a cell culture state when cell growth is not present or to stress events such as in elicitation procedures. The production of secondary metabolites is often produced in two-phase cultures, the first phase supporting cell growth, and the second acting as a metabolic sink for the accumulation of the secondary metabolites in question (Andrews and Roberts 2017).

Perhaps the most important factor influencing the production of secondary metabolites in plants is morphological cell differentiation, which results in the generation of tissues and organs with differentiated cells in terms of metabolic activity and structural organization. Indeed, secondary metabolites are often found in specialized tissues such as flowers, roots, bark, leaves, resin canals, and, in particular, cells within tissues such as laticifers. Because cell differentiation is greatly restricted (and therefore usually considered absent) in suspension cultures, only a limited number of these cultures can produce the required secondary metabolites, even though the plant from which the cell line was generated is able to produce these metabolites. For instance, *in vivo* synthesis of vincristine and vinblastine requires a complex process involving several organelles in specialized cells of aerial tissues (Murata and Luca 2005). Therefore, it has been widely believed that *in vitro* cultures are unable to produce vincristine and vinblastine (Verpoorte et al. 1993). More recently, several studies have refuted the former idea by achieving production in calli with early differentiation (Taha et al. 2014; Ataei-Azimi et al. 2018). Unfortunately, analysis of the role of cell differentiation required for secondary metabolite production has received little attention. Furthermore, suspension cultures producing vincristine and vinblastine have also been reported (Taha et al. 2014), although no explication was given. It is reported that *in vitro* synthesis of the precursor vindoline is the main challenge to overcome since *in vivo* synthesis occurs within the thylakoids of autotrophic cells. Indeed, it has been reported that cell lines derived from cambial meristem cells can synthesize vincristine and vinblastine from vindoline, though supplementation of vindoline is required (Zhang et al. 2015). On the other hand, there have been reports of vindoline-producing cultures (Scott et al. 1980; Naaranlahti et al. 1989). Hence, it is possible that there may be alternative metabolic pathways active in cells lacking autotrophic metabolism, which could be active in suspension cultures (Zavala-Ortiz et al. 2021).

4.7 INNOVATIVE APPROACH: CELL DIFFERENTIATION

Suspension cultures are generally initiated from undifferentiated callus. Resulting cell lines are usually thought to be composed of "undifferentiated" cells, lacking specialized secondary metabolism, as cell differentiation is absent or negligible, even if the cell lines were generated from specialized *in vivo* tissues producing the secondary metabolites (Servet Kefi 2018). It must be clearly stated that the absence of a secondary metabolite implies null expression of its biosynthetic pathway rather than the absence of the related genes. The gene activation during cell differentiation in suspension cultures, could be the key for the establishment of innovative plant metabolites production processes using bioreactors and other techniques of genetic and metabolic engineering to regulate metabolic pathways (Servet Kefi 2018).

Suspension cultures are described as "undifferentiated cultures" largely (and likely dogmatically) in the literature, whereas we hold the view that cell differentiation can be achieved within aggregates of suspension cultures. So, can suspension cultures be composed of differentiated cells or not? There has been some hesitation about the formality of the term "cell differentiation" used in suspension cultures. Therefore, analysis of terms like "undifferentiated cell" and "cell differentiation" must firstly be addressed.

In vivo, cell differentiation is usually considered to be a discrete characteristic depending on the presence of intracellular structures. For instance, depending on the nature of the cell wall, a cell can be classified as a parenchyma, collenchyma, or sclerenchyma cell, such cell types being different in terms of cell wall characteristics, and thus there is differentiation between them (Mauseth 2017). However, the term "cell differentiation" is rarely used in *in vitro* studies and the terms "undifferentiated" or "dedifferentiated" cells are preferred. The terms "undifferentiation" or "dedifferentiation", have received many definitions since the early 1940s, describing a process where cells lose their specialization (Bloch 1941). More recently, the term has been associated with totipotency and therefore defined as the loss of the cell's differentiated state to become a stem-cell-like state, conferring pluripotency (Grafi 2004). Consequently, an undifferentiated cell should likely be a stem cell. However, not all *in vitro* cultures are pluripotent, so the generic use of "dedifferentiated cells" seems inappropriate.

Nowadays, cell dedifferentiation is increasingly referred to as the increase in the developmental potency of cells (Fehér 2019) rather than lack of difference between cells in a biological unit, which has been associated *in vivo* with cell differentiation. Though we partially agree with the former definition, for *in vitro* cultures, cell differentiation should be defined as the process whereby cells become different (independent of its impact on cell pluripotency) from adjacent cells in terms of their structure (shape, content) and physiological state (expression of particular genes, metabolic rates), among others. Then, in our own definition, a "dedifferentiated cell" is a cell that became similar to adjacent cells and could only exist in a homogeneous population where all cells perform in the same way. In brief, the process of "dedifferentiation" is not linked to totipotency or specialization. In addition, that "dedifferentiation" is unlikely to happen, with exceptions, such as in relatively homogeneous synchronous cell culture using a membrane-elution method (Cooper 2002), a "baby plant cell machine" which has not been reported yet. The homogeneous synchronous culture is composed of cells with, statistically speaking, the same structure (DNA, compartmentalization, size, and shape), physiological state (DNA expression state, energetic level), and biological state (all cells being viable, same cell age) (Cooper 2002). Thus, no significant difference between cells is expected in this ideal case.

Cell differentiation, as defined by us, could then lead to the possibility of cell differentiation occurring even in suspension cultures of single cells, provided that *significant* differences exist between cells. So, what is a significant difference? Cell differences in suspension cultures must be related to all processes by which cells become significantly different from each other through unequal expression of the same genetic material for the synthesis of secondary metabolites (Luckner et al. 1977). This concept is extremely broad as it can imply dynamics of intra-, extra-, and inter-cellular mechanisms of differential gene expression, including the description of the origin of morphological and physiological changes between cells (Stange 1965). Consequently, most works have focused only on the detection of direct cell differences rather than on the mechanisms explaining such differences as reviewed in the following sections.

Detection of significant differences between cells within a single culture can become a challenging task, depending on the measured difference. It has been previously mentioned that there can be suspension cultures producing secondary metabolites with no obvious cell differentiation (Taha et al. 2014). These processes are often grouped under the "produced when cell growth is arrested" perspective rather than by the "produced by specialized cells" perspective. However, the two approaches are not mutually exclusive but complementary. There could be cells with different physiological states producing secondary metabolites and "hiding" in the whole cell population as no proper analysis is usually performed to detect such cell differences at the single-cell level. For instance, it was recently discovered that approximately 65% of *Taxus cuspidata* cells are non-cycling during batch culture. These non-cycling cells probably appear similar under the light microscope and it needed the bromodeoxyuridine test and a flow-cytometer for the detection of subpopulations with different cycling capacities (Naill and Roberts 2005). This example demonstrated that even cells similar in terms of size and shape can be extremely different in biochemical terms, namely the physiological state of cells (gene expression, energetic level, metabolic rates, among others). Thus, orthogonal analysis of cell differentiation is required to detect any possible cell differences, particularly at the single-cell level using flow cytometry and cell sorting (Naill and Roberts 2005).

As far as can be ascertained, no consensus has been achieved for the convincing identification of cells with different states or specialization in suspension cultures, though it has been widely stated that cell differentiation is favorable for the production of metabolites of interest (Ratnadewi 2017). Differences in the size and form of cells have been related to different alkaloid synthesis capacities in *Catharanthus* suspension cultures (Deus-Neumann and Zenk 1984), particularly associated with the loss of alkaloid production. Then, the relationship between the morphology of cells and alkaloid synthesis in suspension cultures was investigated. Through the analysis of 11 cell lines, it was discovered that there were two main cell morphologies, either spherical or cylindrical. These shapes were characterized by the aspect cell ratio (major cell axis/minor cell axis). Analysis of the indole alkaloid content of these cells revealed that alkaloid synthesis, particularly of ajmalicine and catharanthine, was significantly higher when the cell aspect ratio (width/length) was greater than 2.8 (Kim et al. 1994). Efforts to identify the sources of these different cell states were performed. The effects of auxins or cytokinins were evaluated, but the plant growth regulators did not affect cell differentiation in terms of cell shape. Finally, cell differentiation was inferred to be "internally determined". Some few reports of cell differentiation in suspension cultures have shown the presence of cells with different cell wall composition (Zavala-Ortiz et al. 2021; Oda et al. 2005; Ogita et al. 2012; Twumasi et al. 2009; Ménard et al. 2017), particularly to study xylogenesis. For instance, in suspension cultures of *Cinchona ledgeriana*, tracheary elements (sclerenchyma-like cells) were observed at the periphery of cell aggregates containing more than 90% of the total alkaloids in the cell culture (Hoekstra et al. 1990). More recently, in preliminary studies, it has been shown that there can be several types of cells within cell aggregates of suspension cultures (Figure 4.2), where tracheary elements (Figure 4.2F) were likely related to alkaloid accumulation in suspension cultures of *Catharanthus* (Zavala-Ortiz et al. 2021). Unfortunately, sclerenchyma differentiation in suspension cultures, particularly into tracheary elements, have mainly focused on ligneous issues and little is known about alkaloid production.

Other reports have used biochemical properties of cells to depict a cell differentiation state, such as variations in vacuole pH, and vacuole color due to the presence of pigments, alkaloids or other secondary metabolites.

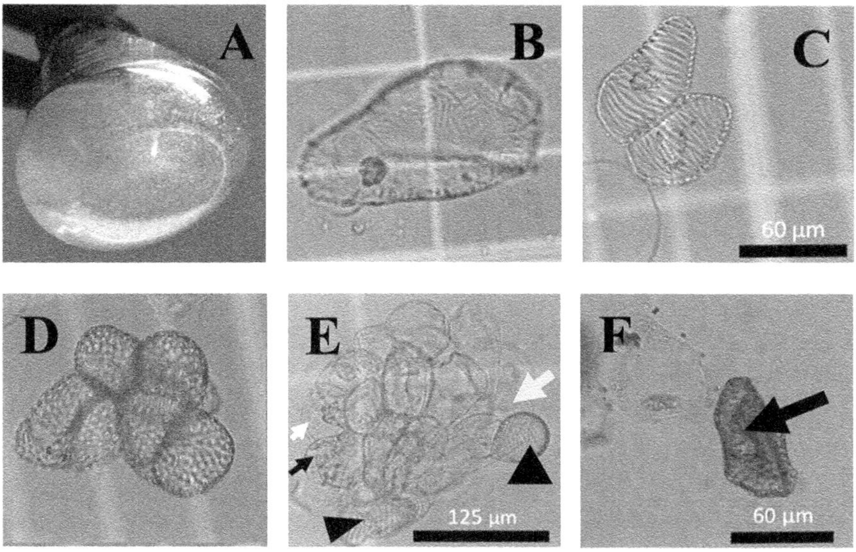

FIGURE 4.2 Cell differentiation occurring in suspension cultures of *Catharanthus roseus*. Suspension culture in 250 mL flask (A). Start of cell differentiation into sclerenchyma-like cell (lignin fibers developing in the primary cell wall) (B). Advanced process of cell differentiation into sclerenchyma-like cell with obvious formation of secondary walls (C). Aggregate of sclerenchyma-like cells (tracheary elements) (D). Typical aggregate containing parenchyma-like cells (white arrows), collenchyma-like cells (dark arrow), and sclerenchyma-like cells (triangles) (E). A sclerenchyma-like cell (arrow) positive for alkaloid content detected with Wagner's reagent and parenchyma-like cells negative for alkaloids (colorless) (F).

For instance, in suspension cultures of *Catharanthus*, it has been reported that there exists a marked heterogeneity of cell subpopulations related to the accumulation of anthocyanins and serpentine (Knobloch et al. 1982). Cell differences were obvious, using light microscopy for the identification of red cells containing anthocyanins and using fluorescence microscopy for the detection of cells containing serpentine. These two types represented approximately 10% of the entire cell population. In this research, some evidence was presented about the factors inducing anthocyanin and serpentine accumulation in different cells. Light exposure, low phosphate and nitrogen availability, and high sucrose concentration were shown to be beneficial for alkaloid accumulation. Unfortunately, no data were provided showing the dynamics of cell differentiation and the eventual need for other types of cells for the accumulation of alkaloids.

In *Catharanthus* suspension cultures, cells with different states were identified by different pH values within their vacuoles (Neumann et al. 1983). By using fluorescence and electron microscopy, it was discovered that alkaloid accumulation occurred inside the vacuoles of particular cells. These alkaloid-accumulating cells exhibited a vacuolar pH of 3, whereas most cells had a vacuolar pH of about 5. Moreover, it seemed that the presence of such alkaloid-accumulating cells was a prerequisite for the accumulation of alkaloids in the suspension cultures. The differentiation into these alkaloid-accumulating cells depended on the culture conditions and the cell line. The identification of cell differentiation through biochemical differences must be considered as cells with these differentiated states (different vacuolar pH) showed similar morphology. Indeed, the great majority of images presented in publications of plant suspension cultures tends to show relatively homogeneous cell populations in terms of cell morphology (Zavala-Ortiz et al. 2021). Therefore, the identification of cell differentiation should consider several and orthogonal analyses for the identification of cell subpopulations with particular specializations related to secondary metabolite synthesis.

4.8 IMPLEMENTATION: MANAGEMENT OF CELL DIFFERENTIATION

It has been reviewed that cell differentiation can be important to produce some secondary metabolites. Secondary metabolites titers in suspension cultures are likely enhanced by the presence of cells with different characteristics and interactions between these different cell types. Therefore, it is of great importance to explore the feasibility of novel processes based on the management of cell differentiation. It is likely that control of culture heterogeneities (i.e., cell aggregate size and gradients, among others) inducing significant differences among cells could determine secondary metabolite production to some extent. Such novel processes would require a deep understanding of what is happening in every single cell or cell subpopulation in response to the cell environment, namely the culture conditions of the bioreactor (Craven and Whelan 2015). However, it is not yet clearly understood how cell differentiation could be controlled for the management and optimization of suspension culture performance (Zavala-Ortiz et al. 2020). Consequently, efforts must firstly be carried out to understand the cell differentiation process and its link to secondary metabolite synthesis in suspension cultures. Then, enhanced production approaches, based on the real time control of cell differentiation, could be launched.

There exist three main challenges for the proper monitoring and control of plant cell suspension cultures: 1) The diverse possible critical variables of a physical, chemical, and biological nature that affect cell differentiation; 2) the inherent challenges for their monitoring, such as the instrumentation and calibration of the monitoring procedures; and 3) the need for a deep process understanding of control strategies.

Currently, the most advanced cell culture processes are supported by automatic control strategies using Process Analytical Technology (PAT) for the continuous measurement and control of critical process parameters (CPP) to ensure final process specifications. The core of these applications is the use of in-line probes (Figure 4.1D) for the measurement of the CPP. There exist several types of in-line cell analyzers based on different fundamental phenomena that could serve to monitor particular properties of the cell culture or cells. For instance, capacitance, also referred to as dielectric permittivity, can serve to monitor the capacity of vesicles (i.e., cells) to store charges. On the other hand, in-line analyzers based on vibrational spectroscopy, such as Raman or Near-Infrared Spectroscopy (NIRS), could provide information about the presence of functional chemical groups or the size of particles within the culture medium. All these analyzers provide spectra that must be used, in a predictive manner, to estimate the content of particular characteristics of the plant suspension culture process. In the context of the innovative processes based on differentiated cells, in-line analyzers providing information on cell subpopulations must first be developed. Then, control actions can be executed for building retro-control procedures (Zavala-Ortiz et al. 2022) in the new proposed processes of differentiated cells.

Although cell cultures are nowadays monitored in terms of several critical variables, using in-line probes (Claßen et al. 2017), they are mostly restricted to physical parameters (i.e., temperature, redox potential, dissolved gases, turbidity). Processes can be enhanced if novel critical variables reflecting the presence of cell subpopulations, namely differentiated cells, are monitored (Clementschitsch and Bayer 2006). Flow cytometry is the standard technique used to analyze cell subpopulations. However, the cost of equipment and the labor-intensive sample preprocessing necessary for analysis prevents cell cytometry from currently being a serious alternative for in-line approaches in plant suspension cultures. Dielectric analyzers are good options for in-line monitoring of cells in terms of size, volume, and shape. Recently, monitoring procedures based on permittivity spectroscopy have been successfully established to monitor plant cell biomass in-line as packed cell volume (PCV) and cell dry weight (DW) (Matanguihan et al. 1994; Holland et al. 2013). Unfortunately, these monitoring procedures still provided a "mean value" of the whole cell population rather than a description of subpopulations within the culture.

New developments, such as micro chips, have combined the advantages of cell cytometry for subpopulation analysis with impedance spectroscopy for providing data from such subpopulations in terms of cell size, membrane capacitance, and cytoplasm conductivity (Cheung et al. 2005). Though this novel approach could be extremely useful, it is currently available only for animal cell cultures without cell aggregation. New developments must then be developed for plant suspension cultures. Furthermore, other strategies must be generated for providing data on subpopulation dynamics throughout the plant suspension culture processes.

More recently, NIRS has proven to be a promising technique for the monitoring of cell subpopulations based on different cell wall compositions (Zavala-Ortiz et al. 2020). In this work, off-line measurements of different cells, classified as parenchyma-, collenchyma-, and sclerenchyma-like cells, and in-line NIR spectra captured in 3 L bioreactor cultures were used to generate monitoring procedures based on Partial Least Squares Regression (PLSR). Then, the models were used to generate in-line estimates of the subpopulation distributions. Though this work pointed out that *in situ* NIR spectroscopy is a promising PAT tool to monitor cells with respect to different cell wall compositions, it is preliminary and issues remain to be clarified.

There are several options for the real-monitoring of cells in cell cultures, although their feasibility in plant suspension cultures has received only limited attention. Raman spectroscopy is attracting interest in mammalian cell cultures due to limited interference from water molecules. It has been used for in-line glycosylation analysis of proteins and CPP as culture medium compounds (Li et al. 2018). However, as far as can be ascertained, its use for the monitoring of different plant cells in suspension cultures has not been performed. There are promising results that could imply that Raman spectroscopy could distinguish different cells in suspension cultures. For instance, an off-line Raman procedure has successfully been employed for the detection of five types of leukocytes in human blood samples (Li et al. 2020). Perhaps, in a few years, such technology could be employed *in situ* for the detection of plant cells with different characteristics.

Previous studies have shown the potential of *in situ* probes for the monitoring of different types of cells, depending on their composition, as terms of size, shape, and type of cell wall. However, the identification of cell subpopulations in terms of physiological state and biochemical characteristics is more challenging as it needs to monitor the concentration of key compounds inside single cells. In this context, fluorescence probes could be used to detect emissions at defined wavelengths for the key compounds. For instance, fluorescence probes have been used for *in situ* monitoring of *Eschscholtzia californica*, *Catharanthus* (Hisiger and Jolicoeur 2008), and *Azadirachta indica* (Srivastava, Harsh, and Srivastava 2008) suspension cultures by the detection of endogenous fluorophores such as NAD(P)H, riboflavins, tryptamine, tryptophan, and fluorescent secondary metabolites. Detection of such metabolites provided information on the biochemical nature of cells as a mean value, though no information on subpopulations could be provided.

In brief, most research has been focused on the monitoring of several parameters in the form of mean values, namely of the entire cell population. Novel alternatives, such as in-line micro-flow cytometers, are emerging (Cheung et al. 2005), then probes could be used to analyze cells individually. However, it seems unlike that such approaches will become available soon or be feasible for plant cells. On the other hand, the use of several and orthogonal monitoring approaches seems to be the most feasible technology. For instance, the use of a dual strategy, based in NIRS and dielectric spectroscopies, was successfully applied for the real-time monitoring of viable, dead, and lysed cell subpopulations in real time during Chinese Hamster Ovary (CHO) cell cultures (Courtès et al. 2016). Multivariate analysis of several orthogonal analyzers could then reveal complex relationships to infer the distribution of several cell subpopulations with some certainty. We think this approach will be a viable alternative for the monitoring and control of plant cell suspension cultures in the future. Then, cell subpopulation dynamics could be real-time monitored and controlled through the proper management of the sources of cell heterogeneities.

4.9 CONCLUSIONS

In this chapter, the production of secondary metabolites has been reviewed from a technical perspective. Suspension cultures seem ideal for large-scale supply in major cases although there are technological limitations, mainly because suspension cultures cannot produce important secondary metabolites which require the presence of different specialized cells in plants. Until now, there have been only a few studies showing that cell differentiation could be achieved in suspension cultures to some extent, opening the path for novel processes for secondary metabolite production based on the control of cell differentiation. However, most of such research has been descriptive in nature and there remains limited process understanding. Therefore, more effort must be carried out to achieve a deeper comprehension of cell differentiation and its management as the base of process control strategies. Management of cell differentiation will require real-time monitoring procedures of cell subpopulations, which will likely be developed using PAT analyzers. Finally, such a combination of process knowledge and control of cell differentiation will lead to innovative production processes for a sufficient supply of plant-derived molecules at affordable prices over the next decades.

ACKNOWLEDGEMENTS

The authors acknowledge the economic support of Consejo Veracruzano de Investigación Científica y Desarrollo Tecnológico (COVEICYDET), and Tecnológico Nacional de México (TecNM) for funding, as well as the support of Prof. Lukas Fischer (Charles University, Prague) and emeritus Prof. Robert Verpoorte (Leiden University, The Netherlands) for their collaborative spirit.

REFERENCES

Alam M. M., M. Naeem, M. M. A. Khan, and M. Uddin. 2017. "Vincristine and vinblastine anticancer *Catharanthus* alkaloids: Pharmacological applications and strategies for yield improvement." In *Catharanthus roseus*, edited by M. Naeem, T. Aftab, and M. M. A. Khan, 277–307. Cham: Springer International Publishing. https://doi.org/10.1007/978-3-319-51620-2_11

Andrews G. R., and S. C. Roberts. 2017. "Bioprocess engineering of plant cell suspension cultures." In *Applied Bioengineering*, edited by T. Yoshida, 283–326. Weinheim, Germany: Wiley-VCH Verlag GmbH & Co. KGaA. https://doi.org/10.1002/9783527800599.ch10

Ataei-Azimi A., B. D. Hashemloian, H. Ebrahimzadeh, and A. Majd. 2018. "High *in vitro* production of anti-canceric indole alkaloids from periwinkle (*Catharanthus roseus*) tissue culture." *African Journal of Biotechnology* 7(16): 2834–39.

Barrett J. 2019. "Limited supply of chemotherapy agent that treats several pediatric cancers may lead to rationing of doses." *Pharmacy Times*. https://www.pharmacytimes.com/view/vincristine-drug-shortage-stirs-concern-for-impact-on-pediatric-patients-with-cancer

Blanco-Carcache P. J., E. M. Addo, and A. D. Kinghorn. 2021. "Higher plant sources of cancer chemotherapeutic agents and the potential role of biotechnological approaches for their supply." In *Medicinal Plants*, edited by H. M. Ekiert, K. G. Ramawat, and J. Arora 28: 545–81. Sustainable Development and Biodiversity. Cham: Springer International Publishing. https://doi.org/10.1007/978-3-030-74779-4_17

Bloch R. 1941. "Wound healing in higher plants." *The Botanical Review* 7(2): 110–46. https://doi.org/10.1007/BF02872446

Carvalho Â., K. G. Vanegas, F. Pereira, S. Theobald, and A. Takos. 2019. "Synthetic biology: An overview." *Comprehensive Biotechnology* 3: 659–70. Elsevier. https://doi.org/10.1016/B978-0-444-64046-8.00202-0

Cheung K., S. Gawad, and P. Renaud. 2005. "Impedance spectroscopy flow cytometry: On-chip label-free cell differentiation." *Cytometry Part A* 65A(2): 124–32. https://doi.org/10.1002/cyto.a.20141

Claßen J., F. Aupert, K. F. Reardon, D. Solle, and T. Scheper. 2017. "Spectroscopic sensors for in-line bioprocess monitoring in research and pharmaceutical industrial application." *Analytical and Bioanalytical Chemistry* 409(3): 651–66. https://doi.org/10.1007/s00216-016-0068-x

Clementschitsch F., and K. Bayer. 2006. "Improvement of bioprocess monitoring: Development of novel concepts." *Microbial Cell Factories* 5(1): 19. https://doi.org/10.1186/1475-2859-5-19

Cooper S. 2002. "Minimally disturbed, multicycle, and reproducible synchrony using a eukaryotic baby machine." *BioEssays* 24(6): 499–501. https://doi.org/10.1002/bies.10108

Courtès F., B. Ebel, E. Guédon, and A. Marc. 2016. "A dual near-infrared and dielectric spectroscopies strategy to monitor populations of Chinese hamster ovary cells in bioreactor." *Biotechnology Letters* 38(5): 745–50. https://doi.org/10.1007/s10529-016-2036-0

Craven S., and J. Whelan. 2015. "Process analytical technology and quality-by-design for animal cell culture." In *Animal Cell Culture*, edited by M. Al-Rubeai 9: 647–88. Cell Engineering. Cham: Springer International Publishing. https://doi.org/10.1007/978-3-319-10320-4_21

Deus-Neumann B., and M. Zenk. 1984. "Instability of indole alkaloid production in *Catharanthus roseus* cell suspension cultures." *Planta Medica* 50(5): 427–31. https://doi.org/10.1055/s-2007-969755

Di Renzo G. C., V. Tosto, and V. Tsibizova. 2020. "Progesterone: History, facts, and artifacts." *Best Practice and Research Clinical Obstetrics and Gynaecology* 69: 2–12. https://doi.org/10.1016/j.bpobgyn.2020.07.012

Eibl R., P. Meier, I. Stutz, D. Schildberger, T. Hühn, and D. Eibl. 2018. "Plant cell culture technology in the cosmetics and food industries: Current state and future trends." *Applied Microbiology and Biotechnology* 102(20): 8661–75. https://doi.org/10.1007/s00253-018-9279-8

Fehér A. 2019. "Callus, dedifferentiation, totipotency, somatic embryogenesis: What these terms mean in the era of molecular plant biology?" *Frontiers in Plant Science* 10: 536. https://doi.org/10.3389/fpls.2019.00536

Flores-Bustamante Z. R., F. N. Rivera-Orduña, A. Martínez-Cárdenas, and L. B. Flores-Cotera. 2010. "Microbial paclitaxel: Advances and perspectives." *The Journal of Antibiotics* 63(8): 460–7. https://doi.org/10.1038/ja.2010.83

Fujita Y. 2007. "Industrial production of shikonin and berberine." In *Novartis Foundation Symposia*, edited by G. Bock and J. Marsh, 228–38. Chichester, UK: John Wiley & Sons, Ltd. https://doi.org/10.1002/9780470513651.ch16

Ganem B., and R. R. Franke. 2007. "Paclitaxel from primary taxanes: A perspective on creative invention in organozirconium chemistry." *The Journal of Organic Chemistry* 72(11): 3981–7. https://doi.org/10.1021/jo070129s

Grafi G. 2004. "How cells dedifferentiate: A lesson from plants." *Developmental Biology* 268(1): 1–6. https://doi.org/10.1016/j.ydbio.2003.12.027

Hisiger S., and M. Jolicoeur. 2008. "Plant cell culture monitoring using an *in situ* multiwavelength fluorescence probe." *Biotechnology Progress* 21(2): 580–89. https://doi.org/10.1021/bp049726f

Hoekstra S. S., P. A. A. Harkes, R. Verpoorte, and K. R. Libbenga. 1990. "Effect of auxin on cytodifferentiation and production of quinoline alkaloids in compact globular structures of *Cinchona ledgeriana*." *Plant Cell Reports* 8(10): 571–4. https://doi.org/10.1007/BF00270055

Holland T., D. Blessing, S. Hellwig, and M. Sack. 2013. "The in-line measurement of plant cell biomass using radio frequency impedance spectroscopy as a component of process analytical technology." *Biotechnology Journal* 8(10): 1231–40. https://doi.org/10.1002/biot.201300125

Kaur P., A. Dey, V. Kumar, P. Dwivedi, R. M. Banik, R. Singh, and D. K. Pandey. 2021. "Recent advances and future prospects of indole alkaloids producing endophytes from *Catharanthus roseus*." In *Volatiles and Metabolites of Microbes*, 449–72. Elsevier. https://doi.org/10.1016/B978-0-12-824523-1.00018-3

Kefi S. 2018. "A novel approach for production of colchicine as a plant secondary metabolite by *in vitro* plant cell and tissue cultures." *Journal of Agricultural Science and Technology A* 8(3): 121–28. https://doi.org/10.17265/2161-6256/2018.03.001

Kim S. W., K. H. Jung, S. S. Kwak, and J. R. Liu. 1994. "Relationship between cell morphology and indole alkaloid production in suspension cultures of *Catharanthus roseus*." *Plant Cell Reports* 14(1): 23–26. https://doi.org/10.1007/BF00233292

Knobloch K.-H., G. Bast, and J. Berlin. 1982. "Medium- and light-induced formation of serpentine and anthocyanins in cell suspension cultures of *Catharanthus roseus*." *Phytochemistry* 21(3): 591–94. https://doi.org/10.1016/0031-9422(82)83146-4

Kuriakose G. C., P. P. C. Palem, and C. Jayabaskaran. 2016. "Fungal vincristine from *Eutypella* spp - CrP14 isolated from *Catharanthus roseus* induces apoptosis in human squamous carcinoma cell line-A431." *BMC Complementary and Alternative Medicine* 16(1): 302. https://doi.org/10.1186/s12906-016-1299-2

Lange B. M. 2018. "Commercial-scale tissue culture for the production of plant natural products: Successes, failures and outlook." In *Biotechnology of Natural Products*, edited by W. Schwab, B. M. Lange, and M. Wüst, 189–218.

Cham: Springer International Publishing. https://doi.org/10.1007/978-3-319-67903-7_8

Li M., B. Ebel, F. Chauchard, E. Guédon, and A. Marc. 2018. "Parallel comparison of *in situ* raman and NIR spectroscopies to simultaneously measure multiple variables toward real-time monitoring of CHO cell bioreactor cultures." *Biochemical Engineering Journal* 137: 205–13. https://doi.org/10.1016/j.bej.2018.06.005

Li W., L. Wang, C. Luo, Z. Zhu, J. Ji, L. Pang, and Q. Huang. 2020. "Characteristic of five subpopulation leukocytes in single-cell levels based on partial principal component analysis coupled with Raman spectroscopy." *Applied Spectroscopy* 74(12): 1463–72. https://doi.org/10.1177/0003702820938069

Luckner M., L. Nover, and H. Böhm. 1977. Secondary Metabolism and Cell Differentiation. In *Molecular Biology, Biochemistry and Biophysics*, Vol. 23. Berlin, Heidelberg: Springer. https://doi.org/10.1007/978-3-642-81102-9

Manning R., and S. Karki. 2019. "Policy brief: Economic profitability of the biopharmaceutical industry." *Bates White Economic Consulting*. https://www.bateswhite.com/media/publication/175_Economic%20profitability%20of%20the%20drug%20industry.pdf

Matanguihan R. M., K. B. Konstantinov, and T. Yoshida. 1994. "Dielectric measurement to monitor the growth and the physiological states of biological cells." *Bioprocess Engineering* 11(6): 213–22. https://doi.org/10.1007/BF00387695

Mauseth J. D. 2017. *Botany: An Introduction to Plant Biology*. 6th edition. Burlington, MA: Jones & Bartlett Learning.

McElroy C., and S. Jennewein. 2018. "Taxol® biosynthesis and production: From forests to fermenters." In *Biotechnology of Natural Products*, edited by W. Schwab, B. M. Lange, and M. Wüst, 145–85. Cham: Springer International Publishing. https://doi.org/10.1007/978-3-319-67903-7_7

Ménard D., H. Serk, R. Decou, and E. Pesquet. 2017. "Establishment and utilization of habituated cell suspension cultures for hormone-inducible xylogenesis." In *Xylem: Methods in Molecular Biology*, edited by M. de Lucas and J. P. Etchhells, 1544: 37–57. New York, NY: Springer. https://doi.org/10.1007/978-1-4939-6722-3_4

Murata J., and V. D. Luca. 2005. "Localization of tabersonine 16-hydroxylase and 16-OH tabersonine-16-*O*-methyltransferase to leaf epidermal cells defines them as a major site of precursor biosynthesis in the vindoline pathway in *Catharanthus roseus*: Mapping biosynthesis of *Catharanthus* alkaloids." *The Plant Journal* 44(4): 581–94. https://doi.org/10.1111/j.1365-313X.2005.02557.x

Naaranlahti T., S. Lapinjoki, A. Huhtikangas, L. Toivonen, U. Kurtén, V. Kauppinen, and M. Lounasmaa. 1989. "Mass spectral evidence of the occurrence of vindoline in heterotrophic cultures of *Catharanthus roseus* cells." *Planta Medica* 55(2): 155–57. https://doi.org/10.1055/s-2006-961911

Naill M. C., and S. C. Roberts. 2005. "Cell cycle analysis of *Taxus* suspension cultures at the single cell level as an indicator of culture heterogeneity." *Biotechnology and Bioengineering* 90(4): 491–500. https://doi.org/10.1002/bit.20446

Neumann D., G. Krauss, M. Hieke, and D. Gröger. 1983. "Indole alkaloid formation and storage in cell suspension cultures of *Catharanthus roseus*." *Planta Medica* 48(5): 20–3. https://doi.org/10.1055/s-2007-969871

Neumann K.-H., A. Kumar, and J. Imani. 2020. *Plant Cell and Tissue Culture: A Tool in Biotechnology: Basics and Application*. 2nd edition. Cham, Switzerland: Springer.

Noble R. L. 1990. "The discovery of the vinca alkaloids-chemotherapeutic agents against cancer." *Biochemistry and Cell Biology* 68(12): 1344–51. https://doi.org/10.1139/o90-197

Oda Y., T. Mimura, and S. Hasezawa. 2005. "Regulation of secondary cell wall development by cortical microtubules during tracheary element differentiation in *Arabidopsis* cell suspensions." *Plant Physiology* 137(3): 1027–36. https://doi.org/10.1104/pp.104.052613

Ogita S., T. Nomura, T. Kishimoto, and Y. Kato. 2012. "A novel xylogenic suspension culture model for exploring lignification in *Phyllostachys* bamboo." *Plant Methods* 8(1): 40. https://doi.org/10.1186/1746-4811-8-40

Palem P. P. C., G. C. Kuriakose, and C. Jayabaskaran. 2015. "An endophytic fungus, *Talaromyces radicus*, isolated from *Catharanthus roseus*, produces vincristine and vinblastine, which induce apoptotic cell death." *PLoS One* 10(12): e0144476. https://doi.org/10.1371/journal.pone.0144476

Palmer E. 2019a. "Pfizer scrambles to fill void after TEVA stops making chemo drug often given to children." *FIERCE Pharma*. https://www.fiercepharma.com/manufacturing/pfizer-scrambles-to-fill-void-after-teva-stops-making-chemo-drug-often-children

Palmer E. 2019b. "TEVA CEO says manufacturing cuts will keep coming even after $3B reduction achieved." *FIERCE Pharma*. https://www.fiercepharma.com/manufacturing/teva-ceo-says-manufacturing-cuts-will-keep-coming-even-after-3b-reduction-achieved

Ratnadewi D. 2017. "Alkaloids in plant cell cultures." In *Alkaloids: Alternatives in Synthesis, Modification and Application*, edited by V. Georgiev and A. Pavlov. InTechOpen. https://doi.org/10.5772/66288

Scott A., H. Mizukami, T. Hirata, and S.-L. Lee. 1980. "Formation of catharanthine, akuammicine and vindoline in *Catharanthus roseus* suspension cells." *Phytochemistry* 19(3): 488–89. https://doi.org/10.1016/0031-9422(80)83216-X

Shao F., I. W. Wilson, and D. Qiu. 2021. "The research progress of taxol in *Taxus*." *Current Pharmaceutical Biotechnology* 22(3): 360–6. https://doi.org/10.2174/1389201021666200621163333

Srivastava S., S. Harsh, and A. K. Srivastava. 2008. "Use of NADH fluorescence measurement for on-line biomass estimation and characterization of metabolic status in bioreactor cultivation of plant cells for azadirachtin (a biopesticide) production." *Process Biochemistry* 43(10): 1121–3. https://doi.org/10.1016/j.procbio.2008.06.008

Stange L. 1965. "Plant cell differentiation." *Annual Review of Plant Physiology* 16(1): 119–40. https://doi.org/10.1146/annurev.pp.16.060165.001003

Suffness M. 1995. "Taxol: Science and Applications." In *Pharmacology and Toxicology*. Boca Raton, FL: CRC Press.

Taha H., K. S. Shams, N. M. Nazif, and M. Seif-El-Nasr. 2014. "*In vitro* studies on Egyptian *Catharanthus roseus* (L.) G. Don V: Impact of stirred reactor physical factors on achievement of cells proliferation and vincristine and vinblastine accumulation." *Research Journal of Pharmaceutical, Biological and Chemical Sciences* 5(2): 330–40.

Twumasi P., J. H. N. Schel, W. van Ieperen, E. Woltering, O. Van Kooten, and A. M. C. Emons. 2009. "Establishing *in vitro Zinnia elegans* cell suspension culture with high tracheary element differentiation." *Cell Biology International* 33(4): 524–33. https://doi.org/10.1016/j.cellbi.2009.01.019

U.S. Government Accountability Office (GAO). 2017. "Drug industry: Profits, research and development spending, and merger and acquisition deals." https://www.gao.gov/products/gao-18-40

Verpoorte R., and A. W. Alfermann. 2000. *Metabolic Engineering of Plant Secondary Metabolism*. Dordrecht: Springer Netherlands. http://public.ebookcentral.proquest.com/choice/publicfullrecord.aspx?p=3106525

Verpoorte R., R. van der Heijden, J. Schripsema, J. H. C. Hoge, and H. J. G. Ten Hoopen. 1993. "Plant cell biotechnology for the production of alkaloids: Present status and prospects." *Journal of Natural Products* 56(2): 186–207. https://doi.org/10.1021/np50092a003

Wilson S. A., and S. C. Roberts. 2012. "Recent advances towards development and commercialization of plant cell culture processes for the synthesis of biomolecules: Development and commercialization of plant cell culture." *Plant Biotechnology Journal* 10(3): 249–68. https://doi.org/10.1111/j.1467-7652.2011.00664.x

Zavala-Ortiz D. A., A. Denner, M. G. Aguilar-Uscanga, A. Marc, B. Ebel, and E. Guedon. 2022. "Comparison of partial least square, artificial neural network, and support vector regressions for real-time monitoring of CHO cell culture processes using *in situ* near-infrared spectroscopy." *Biotechnology and Bioengineering* 119(2): 535–49. https://doi.org/10.1002/bit.27997

Zavala-Ortiz D. A., B. Ebel, E. Guedon, A. Marc, D. M. Barradas-Dermitz, P. M. Hayward-Jones, and M. G. Aguilar-Uscanga. 2020. "*In situ* cell differentiation monitoring of *Catharanthus roseus* suspension culture processes by NIR spectroscopy." *Bioprocess and Biosystems Engineering* 43(4): 747–52. https://doi.org/10.1007/s00449-019-02255-x

Zavala-Ortiz D. A., M. J. Martínez-Montero, E. Guedon, A. Marc, B. Ebel, D. M. Barradas-Dermitz, P. M. Hayward-Jones, M. Mata-Rosas, and M. G. Aguilar-Uscanga. 2021. "Interest of cellular differentiation in the production of vincristine and vinblastine in suspension cultures of *Catharanthus roseus* (L.) G Don." *Revista Mexicana de Ingeniería Química* 20(2): 807–21. https://doi.org/10.24275/rmiq/Bio2228

Zhang W., J. Yang, J. Zi, J. Zhu, L. Song, and R. Yu. 2015. "Effects of adding vindoline and MeJA on production of vincristine and vinblastine, and transcription of their biosynthetic genes in the cultured CMCs of *Catharanthus roseus*." *Natural Product Communications* 10(12): 1934578X1501001. https://doi.org/10.1177/1934578X1501001220

CHAPTER 5

Metabolomics in Medicinal Plants

Lina María Londoño-Giraldo,
Edmundo Lozoya-Gloria, and
Alma Angélica Del Villar-Martínez

CONTENTS

5.1 INTRODUCTION

Since ancient times, plants have been used as medicines, and, currently, around ten percent of the 350,000 species described are used for this purpose (Salmeron-Manzano et al. 2020). Moreover, about one-third of the world's best-selling drugs are obtained from plant natural compounds or their derivatives (Dehelean et al. 2021). In the post-Neolithic period, about 60% of plants were used as medicines; nowadays, people collect wild and cultivated plants for medicinal use, as an indispensable part of human civilization (Hao and Xiao 2015). According to the World Health Organization (WHO), more than 80% of the world's population routinely

DOI: 10.1201/9781003166535-5

uses traditional medicines to meet their primary healthcare needs (Wang et al. 2020). Plants are necessary for the everyday diet and their nutritional values have been intensively studied for decades. In addition, the synthesis of a wide range of secondary metabolites, such as alkaloids, flavonoids, quinones, lignans, steroids, and terpenoids, have been used by humans as pharmaceuticals, agrochemicals, flavors, fragrances, colors, biopesticides, and food additives (Thakur et al. 2019, Leyva-Padrón et al. 2020), representing a huge market for industrial production. Plant secondary metabolites also include a wide variety of elements, such as nitrogen, sulfur, phosphorus, etc. (Leyva-Padrón et al. 2020, Ghassemi et al, 2021). Many of these compounds play an important role in defense mechanisms against bacteria, fungi, viruses, and insects (Reddy et al. 2020, Reyes-Vaquero et al. 2021) (Seca and Pinto 2018, Reddy et al. 2020, Reyes-Vaquero et al. 2021) Research on plant secondary metabolites has increased over the past 50 years, due to their significance in daily life, such as human health care (Jamwal et al. 2018).

Analysis of secondary plant metabolites is particularly important for the understanding of plant physiology, including plant growth, development, and defense mechanisms. However, in the case of herbal medicinal plants (Chua 2016), it is important to study the metabolic interactions when plants or metabolites are used to treat important diseases (Urquiza-López et al. 2021). The production and accumulation of plant metabolites depend on several factors, and it has been proposed that plants assign a huge part of their genomes to regulate, biosynthesize, and store many chemical compounds. These chemicals play a crucial role in the interaction of plants with the environment, and these compounds are known as "specialized metabolites" (Yuan and Grotewold 2020). Secondary metabolite accumulation often occurs in plants in response to different stimuli, such as osmotic stress, temperature, humidity, drought, bacterial, yeast, or fungal extracts, salicylic acid, or physical injury, among others, which are biotic or abiotic elicitors (Naik and Al-Khayri 2016).

Recently, novel studies into the general processes of living systems have developed using omics data (Figure 5.1). These consist of the abundance measurements and structural characterization of a large range of molecules, using different methodologies (Gorrochategui et al. 2016). The chemical structures are involved in important biological processes, and the study of the metabolic pathways includes several omics phases, as follows: genomics, which looks at the DNA to investigate the presence/absence of certain genes; transcriptomics, studying the transcribed genetic material,

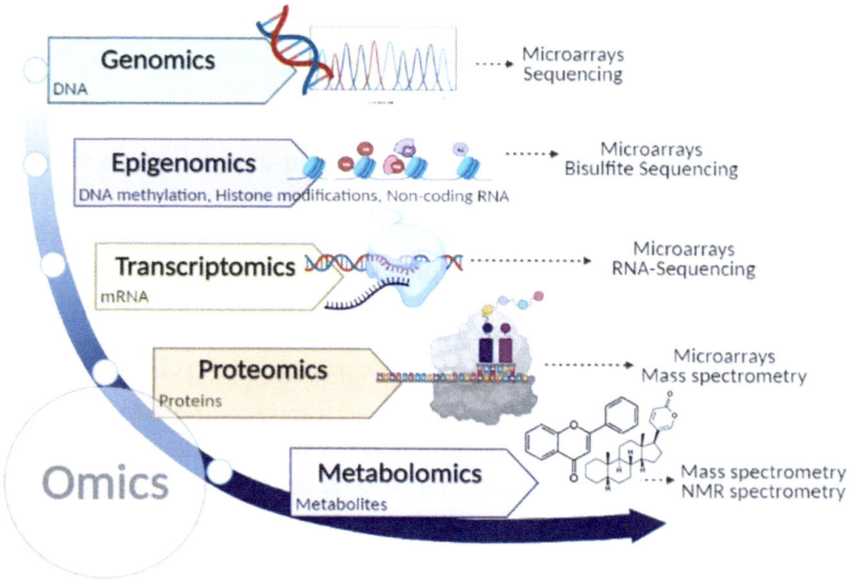

FIGURE 5.1 The omics: visualization of the genetic information flow in plants.

examining the genes which are actively expressed; and proteomics, which helps in characterizing the information flow happening within the cell and the organism, about the functionality of proteins and metabolomics, to provide an invaluable amount of information on the complex network of metabolites in biological systems (Reel et al. 2021).

Metabolomics research is the study of the chemical and biochemical processes of biological systems, assisted by databases. It makes it possible to understand the metabolic processes and reveal the key pathways or mechanisms used in pharmaceutical research and drug treatments. It is part of the complex analysis necessary to reach important conclusions regarding biological systems. This is a journey that describes the new vision that metabolomics offers and explains recent associated advances to successfully exploit plants of pharmaceutical interest.

5.2 METABOLOME AND PLANT METABOLOMICS

The metabolome is complex and dynamic because different chemical reactions are continuously occurring in the cell, resulting in the formation of hundreds of low-molecular-weight metabolites, present within a biological system. It includes different compounds, like amino acids, fatty acids, carbohydrates, vitamins, and lipids among others; the diversity of molecules within the metabolome is specific to the organism studied (Viant et al. 2017, Urquiza-López et al. 2021). Analysis of the plant metabolome is particularly challenging as the complexity of biological systems increases due to the chemical diversity of primary and secondary metabolites. In plants, the metabolome has been estimated to contain approximately 200,000 metabolites (Heyman and Dubery 2016), and comprehensive analysis has been performed by metabolomic technologies (Bingol 2018). These include a multidisciplinary approach combining analytical chemistry, biochemistry, informatics, and biostatistics for analyzing the set of metabolites contained in biological organisms (Fenaille et al. 2017).

Metabolomics is a relatively young discipline compared with other well-established "omics" such as genomics, transcriptomics, and proteomics (Li et al. 2017). It is a valuable tool in different disciplines, allowing the discovery of new medicines, finding biomarkers, studying diseases, identifying metabolic pathways, and conducting comparative statistical analysis between two or more conditions that can be applied in clinical, toxicological, and pharmacological trials (Steuer et al. 2019). Metabolomics applied to agriculture, including food and medicinal production, is a powerful tool for evaluating the natural variation of the metabolite content among different plant species. Metabolite accumulation can be studied during growth and plant development and used to characterize the metabolic response to different biotic or abiotic stresses (Li et al. 2018). However, this technology represents a significant analytical challenge, unlike genomic and proteomic methods, since the studied molecules are variable and with specific physicochemical properties, such as different polarities, from water-soluble organic acids to nonpolar lipids (Clish 2015). Different chemical species can be found in highly variable concentrations, from individual molecules to high molar fractions (Keppler et al. 2018). Recent advances in analytical methods and data analysis have resulted in increased sensitivity, precision, and capacity, allowing the analysis of several hundred or even thousands of compounds within a sample; thereby, metabolomics continues evolving and improving (Clish 2015, Tenenboim and Brotman 2016).

The metabolome is the complete set of cellular metabolites included in an extract from a biological sample. The metabolome reflects the response of an organism to multiple environmental factors, and includes both the endometabolome and exometabolome (Seenivasan et al. 2020). It can be divided into two categories: non-directed metabolomics and targeted metabolomics, the latter including metabolite fingerprinting and metabolic footprinting. It is necessary to take care of a) sample handling, b) the external variables that should influence the metabolome from samples, and c) the analysis and interpretation of the data. The correct transformation of the data and the use of online platforms will not only reveal the possible differences but also reduce the time of analysis. In conclusion, metabolomics is a multidisciplinary science where the metabolic conclusion obtained will depend on the synergy of an appropriate chemical and instrumental analysis, a data transformation, and a biological interpretation appropriate to the nature of the sample or the biological system analyzed.

The non-directed metabolomics, or "untargeted metabolomics", pretends to profile endogenous metabolites without preferences (Beger et al. 2019). This untargeted approach includes the comprehensive analysis of all measurable analytes in a sample, including uncharacterized metabolites (Gorrochategui et al. 2016, Rochat 2016). The main outstanding feature of untargeted metabolomics is the discovery of novel metabolites related to the background research. The detection and identification of metabolites are usually carried out with mass spectrometry (MS) and nuclear magnetic resonance (NMR) (Wang et al. 2019, Bingol 2018, Emwas et al. 2019). Untargeted metabolomic methods are global in scope and have the aim of simultaneously measuring as many metabolites as possible from biological samples without bias, whereas targeted metabolomic approaches are commonly driven by a specific biochemical question or hypothesis that motivates the investigation of a particular pathway (Cai et al. 2015).

Targeted metabolomics focuses on the analysis of a predefined combination of molecules, with preexisting information (Beger et al. 2019). Data analysis in targeted metabolomics aims to process datasets coming from a predefined group of metabolites, chemically characterized and biochemically identified and contained in referential databases (Gorrochategui et al. 2016). Even though this is still a research analysis, it has true potential for clinical applications and routine analysis (Rochat 2016). In targeted analyses, a small number of predefined metabolites, typically fewer than 20 metabolites are analyzed in the samples of interest. It supplies a high level of sensitivity, precision, and accuracy (Hasanpour et al. 2020).

"Metabolite fingerprinting" determines endogenous cellular markers and evaluates the most representative metabolites to find differences between groups and treatments (Keppler et al. 2018). It provides a comprehensive set of features ideally corresponding to all chemical constituents and aims to extract the non-evident chemical information included among the entire set of signals acquired through an analytical instrumental technique (Stilo et al. 2021). This is a powerful approach for the characterization of phenotypes and the differentiation of specific metabolic conditions due to environmental factors (Liang et al. 2015, Urquiza-López et al. 2021).

Currently, fingerprinting has been applied to specific areas of study, such as food (Cifuentes 2017), agriculture (Sherriff et al. 2018), nutrition (Ghanbari et al. 2019), and medicinal plants (Pérez-Mendoza et al. 2020), among others, and the development of specific analytical techniques and informative multidimensional platforms, which are readily available, will offer further possibilities to continue developing this concept (Stilo et al. 2021).

Finally, the exometabolome "metabolic footprinting" focuses on the analysis of the low-molecular-weight metabolites that are excreted by the cells (Bingol 2018). Basically, this is a strategy for analyzing the properties of cells or tissues by looking in a high-throughput manner at the metabolites that are excreted or fail to be taken up from their surroundings. Metabolic footprinting analysis is the qualitative study of extracellular metabolites (exometabolites) and is another promising technique in metabolic engineering. The results from this technique reflect the influence of environmental conditions on intracellular metabolism, which may be studied by simply probing exometabolites.

In summary, metabolic fingerprinting relies on the measurement of intracellular metabolites, whereas metabolic footprinting depends on the monitoring of metabolites consumed from, and secreted into, the growth medium (Seenivasan et al. 2020).

5.3 ANALYTICAL TECHNIQUES AND BIOINFORMATIC TOOLS FOR METABOLOMIC ANALYSIS

The most common platforms used for metabolomics include liquid chromatograph–mass spectrometry (LC–MS), gas chromatography–mass spectrometry (GC–MS), and nuclear magnetic resonance (NMR) (Zhao et al. 2018).

Mass spectrometry (MS) is currently the most popular technique for metabolite detection and identification due to its ability to be combined with gas chromatography (GC–MS), liquid chromatography (LC–MS), and capillary electrophoresis (CE–MS), and to perform tandem experiments (MS/MS), to enhance sensitivity (Rathahao-Paris et al. 2016, Chen et al. 2018). LC–MS currently stands as the most versatile and commonly used platform for profiling extracts (Allard et al. 2017). Diverse types of

MS have been hyphenated with LC systems, which can be divided into low-resolution (LR) MS (like quadrupole (Q), triple quadrupole (QQQ) and ion trap (IT)) and high-resolution (HR) MS (like time-of-flight (TOF), Q-TOF and Orbitrap) (Yan and Xu 2018).

Various chemical changes can be performed on an analyte in order to make it suitable for a specific method of analysis, the most common being derivatization (Moldoveanu and David 2019). Derivatization is based on converting compounds into a derivative that can be better be analyzed; usually, this conversion is specific for a functional group or a class of functional groups (Holländer 2017). Active hydrogens in the functional groups of molecules containing carboxylic acids (–COOH), alcohols (–OH), amines (–NH_2), and thiols (–SH) can be derivatized by alkylation, acylation, or silylation (Beale et al. 2018).

Metabolomic profiling by NMR has often been performed by one-dimensional (1D) ^{1}H-NMR experiments. The advantage of 1D ^{1}H-NMR is the short acquisition time, which is typically less than one hour/sample, and high sensitivity; it can detect metabolites down to single-digit μM concentrations (Bingol 2018).

Metabolite identification is one of the most difficult challenges in metabolomics, for which spectral knowledge databases are essential. There are databases for the identification of metabolites, databases for the analysis of the biosynthetic pathway, and tools for the spectral process and data analysis (Arnald et al. 2015). These databases are METLIN, Kyoto Encyclopedia of Genes and Genomes (KEGG), Metabolic Pathway Database (MetaCyc), BioCyc Database Collection, MetaboAnalyst, Metabox, OmicsNet, MetScape3, and MetExplore (Table 5.1).

METLIN (https://metlin.scripps.edu/landing_page.php?pgcontent=mainPage), a freely accessible cloud-based technology platform and metabolite database, has grown from a small collection of MS/MS spectra (Guijas et al. 2018). The database contains over 240,000 compounds, which include endogenous metabolites from different organisms and exogenous compounds such as pharmaceutical drugs and other synthetic organic compounds (Levin et al. 2016). METLIN data are broadly useful across multiple tandem mass spectrometry instrument types, with the data collected in both positive and negative ionization modes at multiple collision energies, supplying high-resolution spectra, systematically bought, and manually curated directly from standards and their stable isotope analogs. These data complement the information from other databases,

TABLE 5.1 Databases for the Identification of Metabolites, Analysis of the Biosynthesis Pathway, and Tools for Spectral Process and Data Analysis

Database	Website	Reference
Identification of metabolites		
METLIN	https://metlin.scripps.edu	Guijas et al. (2018)
Analysis of pathways and biosynthesis		
Kyoto Encyclopedia of Genes and Genomes (KEGG)	https://www.genome.jp/kegg/	Kanehisa et al. (2017)
Metabolic Pathway Database (MetaCyc)	https://metacyc.org/	Caspi et al. (2018)
BioCyc Database Collection	https://biocyc.org/	Karp et al. (2019)
Data analysis, interpretation, and integration with other omics data		
MetaboAnalyst	https://www.metaboanalyst.ca/	Chong and Xia (2018)
Metabox	http://kwanjeeraw.github.io/metabox/	Wanichthanarak et al. (2017)
Omicsnet	https://www.omicsnet.ca/home.xhtml	Zhou and Xia (2018)
MetExplore	https://metexplore.toulouse.inrae.fr/index.html/	Cottret et al. (2018)

which have been collected for electron impact (EI) or nuclear magnetic resonance (NMR) instrumentation (Guijas et al. 2018).

KEGG consists of four databases of pathways, genes, compounds, and enzymes. KEGG (https://www.genome.jp/kegg/) is an integrated database resource consisting of fifteen manually verified databases and a computationally generated database in four categories: Systems Information, Genomic Information, Chemical Information, and Health Information (Kanehisa et al. 2017).

MetaCyc (https://metacyc.org/) is a highly supervised database for over 20 years reference to metabolism from all domains of life. It has data about chemical compounds, reactions, enzymes, and metabolic pathways that have been experimentally confirmed and reported in the scientific literature (Caspi et al. 2018).

BioCyc (https://biocyc.org/) is a microbial genome web portal that combines thousands of genomes, with other information inferred by computer programs, imported from other databases (DBs), and validated from the biomedical literature by biologist curators. BioCyc also supplies an extensive range of query tools, visualization services, and analysis software (Karp et al. 2019).

MetaboAnalyst (https://www.metaboanalyst.ca/) is an easy-to-use web-based tool suitable that allows users to perform a wide variety of metabolomic data analysis tasks. The current release (v4.0) holds twelve modules that can be placed into four categories: exploratory statistical analysis, functional analysis, data integration, and data processing (Chong and Xia 2018).

Metabox (http://kwanjeeraw.github.io/metabox/) is a bioinformatics toolbox for deep phenotyping analytics that combines data processing, statistical analysis, functional analysis, and integrative exploration of metabolomic data within proteomic and transcriptomic contexts (Wanichthanarak et al. 2017).

OmicsNet, (https://www.omicsnet.ca/) is a novel tool that allows users to easily create distinct types of molecular interaction networks and visually explore them in a three-dimensional (3D) space. Users can upload one or multiple lists of molecules of interest (genes/proteins, microRNAs, transcription factors, or metabolites), to create and merge several types of biological networks (Zhou and Xia 2018).

MetScape3 (http://metscape.ncibi.org/) and MetExplore (https://metexplore.toulouse.inrae.fr/index.html/) offer an all-in-one online solution composed of interactive tools for metabolic network curation, network exploration, and omics data analysis (Cottret et al. 2018).

5.4 METABOLOMICS APPLIED TO MEDICINAL PLANTS

Medicinal plants are a huge source of natural organic compounds with promising prospects for pharmaceutical applications. Compared with animals and microorganisms, plants can produce far more bioactive metabolites, such as alkaloids, flavonoids, quinines, and terpenoids, and many of them have been confirmed to possess medicinal value (Ma et al. 2019, Leyva-Padrón et al. 2021, Urquiza-López et al. 2021). Atanas et al. (2015) summarized some plant-derived natural products approved for therapeutic use in the period 1984–2014. These products have a diverse range of molecular targets and have been used for the treatment of diverse diseases. Metabolomics has been important for the characterization of extracts; this strategy, accompanied by complementary tests, has allowed proposing the chemical compounds responsible for biological activity. Some examples, such as *Capsicum annuum,* is used externally to treat rheumatism, lumbago, and neuralgia (Srinivasan 2016). Capsaicin, the main metabolite from *C. annuum* fruits, has been studied clinically as a topical treatment for the pain of rheumatoid and osteoarthritis (Persson et al. 2018), psoriasis, diabetic neuropathy, and postherpetic neuralgia (Srinivasan 2016, Hernández-Perez et al. 2020).

In *Cannabis sativa,* some 100 cannabinoids have been identified so far. The main types of natural cannabinoids belong to the families of the cannabigerol, cannabichromene, cannabidiol, cannabinodiol, tetrahydrocannabinol, cannabinol, cannabitriol, cannabielsoin, isocannabinoids, cannabicyclol, cannabicitran, and cannabichromanone types. The most abundant is *trans*-Δ-9-tetrahydrocannabinol (THC), which has been used as an anti vomiting drug in cancer chemotherapy and as an appetite stimulant, especially for AIDS patients. Cannabidiol (CBD) and cannabinol (CBN), with no psychotropic effect, cannabichromene and related compounds possess anti-inflammatory, anti fungal, and anti microbial activities (Gonçalves et al. 2019). *Galanthus caucasicus* and *Galanthus nivalis* produce galanthamine, a morphine-like alkaloid. It has been used as a possible therapeutic agent in Alzheimer's disease because of its central cholinergic effects (Tsakadze et al. 1969, Khonakdari et al. 2018). Metabolomic studies can also help to investigate the pharmacological effects and the action mechanism of a range of herbs (Hasanpour et al. 2020).

Metabolites in medicinal plant research results in thousands of scientific papers yearly. The last five years have been influenced mainly by studies generated by China and India, with 800–1,100 publications per year (Salmonerón-Manzano et al. 2020). In a recent bibliometric study, Salmonerón-Manzano et al. (2020) mentioned that *Artemisia annua*, *Aloe vera*, *Panax ginseng*, *Punica granatum*, *Apocynum cannabinum*, and *Andrographis paniculate* are the most studied medicinal plants. Table 5.2 summarizes some papers that include the development of metabolomic research strategies in several locations of the world; this is the case for *Rosmarinus officinalis*, *Camellia sinensis*, *Valeriana officinalis*, *Melissa officinalis*, and *Salvia officinalis* commonly used in Asia and the Americas for multiple diseases. Asian medicinal plants recorded in recent years in databases include metabolomic approaches to medicinal plants associated with skin care, management of diabetes, cancer treatments, and cardiovascular diseases (Salmonerón-Manzano et al. 2020).

5.5 WORLD TRENDS ON METABOLOMICS IN MEDICINAL PLANTS

People in different parts of the world use, commercialize, and explore diverse plants for medicinal issues, and there has been an increase in investigations that now include metabolomic techniques in specific subjects.

TABLE 5.2 Metabolomics of Popular Medicinal Plants, Medicinal Action, and the Analytical Platform Used Since 2020

Medicinal Plant	Major Analytical Platform	Potential Medicinal Action	Reference
Lupinus albus L.	(^{1}H-NMR)[1] UHPLC-ESI-MS/MS[2]	• Diabetes	Hellal et al. (2021)
Muntingia calabura L.	(^{1}H-NMR)	• Gastroprotection	Zolkeflee et al. (2021)
Salvia officinalis L	LC-ESI/LTQ-Orbitrap/MS[3]	• Inflammatory diseases	Samani et al. (2021)
Cannabis sativa L.	GC–MS[4]	• Cancer • Arthritis	Rashid et al. (2021)
Peganum harmala L. *Zygophyllum album* L.f. *Anacyclus valentinus* L. *Ammodaucus leucotrichus* Coss. & Durand *Lupinus albus* L. *Marrubium vulgare* L.	(^{1}H-NMR)	• Antidiabetic	Hellal et al. (2020)
Lonicerae japonicae Thunb.	LC–QTOF/MS[5]	• Anti-inflammatory	Cai et al. (2020)
Ocimum sanctum L. *Ocimum gratissimum* (L.) *Ocimum kilimandscharicum* Gürke	GC–MS	• Hypoglycemic effect	Rastogi et al. (2020)
Cinnamomum camphora (L.) J.Presl	LC–QTOF/MS	• Sedative • Antispasmodic • Diaphoretic • Anthelmintic	Yang et al. (2020)
Salvia officinalis L.	LC–QTOF/MS	• Antiseptic • Anti-inflammatory	Sharma et al. (2020)
Burkea africana Hook	(^{1}H-NMR)	• Cancer	Nemadodzi et al. (2020)
Catharanthus roseus (L.) G.Don	UHPLC-MS/MS[6]	• Leukemia • Hodgkin's lymphoma	Fraser et al. (2020)
Rosmarinus officinalis L. *Lavandula officinalis* L. *Matricaria chamomilla* L. *Camellia sinensis* (L.) Kuntze *Pelargonium graveolens* L'Hér.	UPLC-HR-ESI-MS/MS[7]	• Anti-aging • Skin protective	Salem et al. (2020)

(*Continued*)

TABLE 5.2 (CONTINUED) Metabolomics of Popular Medicinal Plants, Medicinal Action, and the Analytical Platform Used Since 2020

Medicinal Plant	Major Analytical Platform	Potential Medicinal Action	Reference
Valeriana officinalis L. *Melissa officinalis* L. *Hypericum perforatum* L. *Passiflora incarnata* L.	GC–MS LC–QTOF/MS	• Sedative • Anxiolytic • Sleep disorders	Gonulalan et al. (2020)

[1] Proton nuclear magnetic resonance spectroscopy
[2] Ultrahigh-performance-liquid chromatography–electrospray ionization tandem mass spectrometry
[3] Liquid Chromatography coupled with high-resolution mass spectrometry
[4] Gas chromatography–mass spectrometry (GC–MS)
[5] Ultrahigh-performance-liquid chromatography–Quadrupole Time of Flight–mass spectrometry.
[6] Ultra-performance liquid chromatography–tandem mass spectrometry

The world trends with bibliographic databases and bibliometric tools, using the words "metabolomics studies" in "medicinal plants" (TITLE-ABS-KEY (medicinal AND plants AND metabolomics)). According to the SCOPUS database, until March 2021, has shown different conclusions. Moreover, the number of publications in indexed journals between 2013 and 2020 has increased. China is the leading country of research on the metabolomics of medicinal plants, followed by the United States, India, and Germany. When we focus on the analysis of metabolomics applied to medicinal plants, the nodes network is generated according to keywords co-occurrence (Figure 5.2, graph realized in VOSviewer version 1.6.16) (Perianes-Rodriguez et al. 2016) and allows the differentiation of four nodes that match the main themes studied under this topic. The first node, in red, at left is associated with the metabolomic analysis of the effects of metabolites from plants on animals or human beings. The second node, in green at the upperside, is related to papers that associate metabolomics with genomics, proteomics, production of medicines, and analysis of metabolites of special traits as flavonoids or alkaloids. The third node, in blue at the downside, is based on analytical platforms used for metabolomics in medicinal plants and statistical methodologies for data analysis and interpretation. This approach is also highlighted by Okada et al. (2010). These authors affirmed that score plot, loading plot analyses, and discriminant map analyses are useful for the reduction of a metabolite fingerprint and the classification of analyzed samples. Finally, the fourth

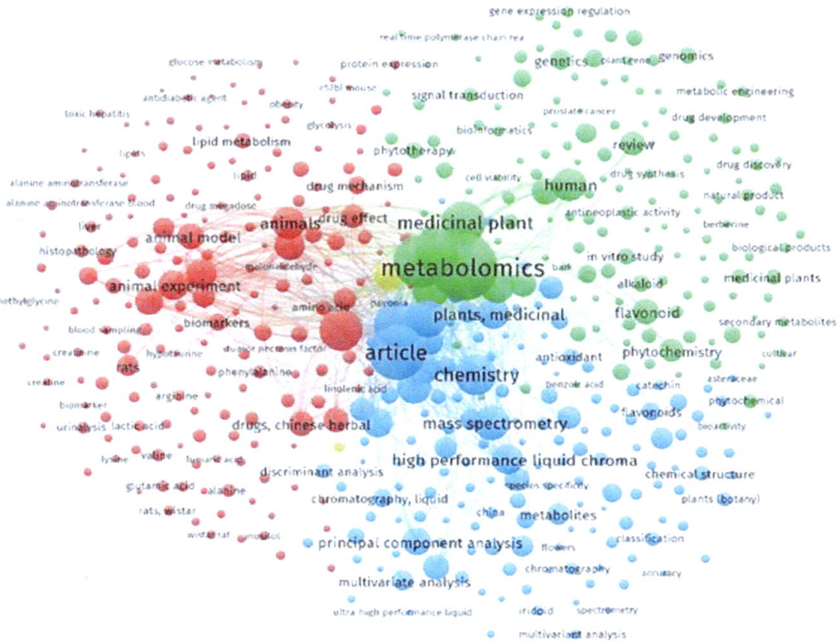

FIGURE 5.2 Bibliometric trends of research in metabolomics in medicinal plants according to SCOPUS exploration. Node network is based on co-occurrences of keywords medicinal (using medicinal AND plants AND metabolomics) of papers available in SCOPUS database.

node, in yellow in the center, is for connection to metabolomics in medicinal plants and Chinese herbal uses.

Metabolomic studies are applied worldwide by the scientific community to explore the chemical profiles and metabolic pathways of various biofluids/herbal extracts (Mumtaz et al. 2017). In synthesis, the revolution of metabolomic approaches to medicinal plants was prompted by the development of analytical platforms, and obviously by the continuous improvement of databases and bioinformatics tools that help researchers to analyze more information in less time.

Different analytical platforms and, in some cases, combinations of techniques are often used as tools for different aims in plant metabolomics. Applying targeted metabolomics, the analysis of specific metabolites in some species or samples is usually combined with enzymatic activity assays. Hussin et al. (2019) used ^{1}H-NMR for analyzing some plants with anti aging effects, such as *Curcuma longa*, *Oenanthe javanica*, *Vitex negundo*, *Pluchea indica*, *Cosmos caudatus*, and *Persicaria minus*, coupled with multivariate data analysis (MVDA) and the free radical scavenging activity of the extract. The results showed the presence of higher levels of metabolites in *P. minus*, such as quercetin, quercetin-3-*O*-rhamnoside (quercitrin), myricetin derivatives, catechin, isorhamnetin, astragalin, and apigenin, having the highest radical scavenging activities and anti-aging properties. Murugesu et al. (2019) reported the metabolomics approach supporting a metabolite profiling to characterize α-glucosidase inhibitors from *Clinacanthus nutans* through LC-MS-QTOF and the partial least square (PLS) statistical model. Four α-glucosidase inhibitors were profiled: (1) 4,6,8-megastigmatrien-3-one; (2) *N*-isobutyl-2-nonen-6,8-diynamide; (3) 1′,2′-bis(acetyloxy)-3′,4′-didehydro-2′-hydro-β, ψ-carotene; and (4) 22-acetate-3-hydroxy-21-(6-methyl-2,4-octadienoate)-olean-12-en-28-oic acid. In addition, molecular docking was used to explain the mechanisms of the interactions involved between the inhibitors and enzymes for anti-α-glucosidase action. Khoo et al. (2015) used ^{1}H-NMR and UHPLC-MS/MS, coupled with PLS statistical modeling, for an untargeted analysis in leaves and stems of *Clinacanthus nutans*. In this study, the metabolite profile was correlated with the total phenolic content (TPC), antioxidant, and α-glucosidase inhibitory activities. Authors reported several new compounds in this plant, including gendarucin A, a gendarucin A isomer, 3,3-di-*O*-methylellagic acid, ascorbic acid, and two isomeric oxoprolinates. The metabolomic analysis also showed that air-dried leaf extracts showed more antioxidant capacity and α-glucosidase inhibitory activities than the stem extracts.

Biomarkers are substances produced by a biological system under specific conditions. Specific molecules can be determined to reflect normal biological processes, pathogenic processes, or as a response to a stimulus or specific treatment (Califf 2018). Particularly in plants, it is useful to analyze the abundance of chemical compounds at different stages of development. Javadi et al. (2015) reported an untargeted study, to investigate the metabolomic changes and the identification of biomarkers during post-harvest storage of *Cosmos caudatus*. The metabolites were identified by gas chromatography–mass spectrometry (GC–MS). Metabolites such as α-tocopherol, catechin, cyclohexen-1-carboxylic acid, benzoic acid, myo-inositol, stigmasterol, and lycopene were found at early storage stages; however, sugars such as sucrose, α-D-galactopyranose, and furanose were detected as biomarkers at late storage stages.

Diseases such as diabetes, cancer, cardiovascular diseases, and anti-inflammatory processes have been considered in metabolomics papers related to medicinal plants. Diabetes is a condition attracting worldwide attention. Hasanpour et al. (2020) summarized the metabolomics findings of this condition, suggesting potential biomarkers and characterizing the metabolic disturbances associated with diabetes development. Guo et al. (2019) reviewed studies on metabolism-targeting anticancer therapies based on metabolomics, with emphasis on the potential of a treatment with pure compounds, herb extracts, and formulations from Chinese medicines, including several metabolites as promising candidates as effective neoadjuvants for cancer treatment. Mir et al. (2020), in parallel, highlighted the importance of research in cancer treatment and medicinal plants, supplying an alternative to existing approaches.

Currently, COVID-19 is a disease caused by a coronavirus-type virus that has been declared the cause of a world pandemic, resulting in an alarming number of deaths, especially among vulnerable citizens around the world. Although several therapeutic molecules are being tested and vaccination is in process, some countries have a long delay before widespread vaccination can be carried out and so need to apply different strategies to attenuate the symptoms. The bioavailability of natural products with putative anti-SARS-CoV-2 effects, such as tannins, should be considered, along with the need for clinical validation of their usefulness and safety (Bernaba et al. 2020). Some studies are suggesting the use of medicinal plants or natural products, either used alone or in combination, as alternative medicines to treat/prevent COVID-19 infection. In China, the use of "Lianhuaqingwen" (LHQW) is used to treat respiratory tract infectious diseases, viral infections, inflammations, and fever, as an herbal mixture of more than ten plants. This mixture (LHQW) could help to treat influenza A infection by improving pathologic alterations, reducing virus replication and lung lesions, as well as inflammation (Runfeng et al. 2020). Through metabolomic analysis of treated serum from mice, two important metabolites were identified, prostaglandin F2α and arachidonic acid, as vital indicators of LHQW in treating influenza (Gao et al. 2020). Jia et al. (2015), using Ultra-Performance Liquid Chromatography method coupled to Diode Array Detection and Quadrupole Time-Of-Flight Mass Spectrometry (UPLC-DAD-QTOF-MS) to analyze the major constituents of LHQW capsules, showed 12 representative compounds as chemical markers, namely salidroside, chlorogenic acid, forsythoside E, cryptochlorogenic acid, amygdalin, sweroside, hyperin, rutin, forsythoside A, phillyrin, rhein, and glycyrrhizic acid. More than sixty natural molecules, originating from 15 medicinal plant species, were evaluated regarding their inhibitory activity toward SARS-CoV helicase (Bernaba et al. 2020). These included myricetin and scutellarein, suggested as promising future anti-SARS drugs (Yu et al. 2012). Qamar et al. (2020), analyzed the 3CLpro (viral 3-chymotrypsin-like cysteine protease) sequence, which controls coronavirus replication and is essential for its life cycle. The 3D homology model was constructed and screened against a medicinal plant library containing 32,297 potential antiviral phytochemicals/traditional Chinese medicinal compounds. These analyses revealed that the top nine hits might serve as potential anti-SARS-CoV-2 lead molecules for further optimization and drug development processes to combat COVID-19. Tripathi et al. (2020) evaluated the potential of 40 natural chemical constituents of *Withania somnifera* to identify a possible inhibitor against the main SARS-CoV-2 protease by a computational approach. The docking study revealed that four constituents of *W. somnifera*, withanoside II, withanoside IV, withanoside V, and sitoindoside IX, exhibited the highest docking energy among the selected natural constituents. Furthermore, a molecular dynamics simulation study of 100 ns predicted that withanoside V may have strong binding affinity and hydrogen-bonding interactions with the protease active site. The putative stability of the active site and their study suggested that withanoside V of *W. somnifera* could be a potential inhibitor against Mpro of SARS-CoV-2 to combat COVID-19 and may have an antiviral effect on novel coronavirus strains. Chikhale et al. (2020) evaluated *W. somnífera*, a plant well known as having antiviral, immunomodulatory, anti-inflammatory, and antioxidant activities, by using molecular docking and dynamics studies. Two different proteins of SARS-CoV-2, namely NSP15 endoribonuclease and the receptor-binding domain of the prefusion spike protein from SARS-CoV-2, were targeted. Molecular docking studies suggested that withanoside X and quercetin glucoside from *W. somnifera* had favorable interactions at the binding site of the selected proteins. The top-ranked phytochemicals from docking studies, subjected to 100 ns molecular dynamics, proposed withanoside X with the highest binding free energy as the most promising inhibitor. The analysis of a mixture of compounds in plant extracts seems quite promising and the development of metabolomics methods based on a small amount of plant material allows the generation of data on the entire spectrum of substances in the sample under study, containing several compounds with antiviral activity (Matveeva et al. 2020). In general, the studies around this disease have been focused on several points, including the angiotensin-converting enzyme II (ACE2), which was found to be a key functional receptor for the SARS-CoV-2, allowing attachment to human and bat cells, and therefore promoting virus replication (Bernaba et al. 2020). Patten et al. (2016) reported 141 medicinal plants and 49 purified natural compounds with documented ACE inhibitory potential. Extracts of *Berberis integerrima*, *Crataegus laevigata.*, *Onopordum acanthium*, and *Quercus infectoria* were evaluated; the last one was found to achieve 94% inhibition of ACE (Shariffi et al. 2013). Finally, and using metabolomic techniques, a natural product named emodin (1,3,8-trihydroxy-6-methyl-anthraquinone) was proposed to inhibit the SARS-CoV – ACE2 interaction (Bernaba et al. 2020). This compound is found in species belonging to the plant families Polygonaceae, Labiatae, Oleaceae, Magnoliaceae, Lauraceae, and Nelumbonaceae, and is produced at high concentrations in the genera *Rheum* and *Polygonum* (Dong et al. 2016).

5.6 PERSPECTIVES

The treatment for or cure of various important diseases or conditions for humanity is still strongly associated with plants of medicinal interest. So, analysis of the entire plant, leaves, roots, stems, flowers, extracts, or of isolated metabolites, is a topic that will always be discussed. Today, research into medicinal and pharmaceutical plants is a topic that demands an interest in national and international journals worldwide, treating different fields on omics with a significant scientific output. In the topic of this chapter, comparative metabolomic platforms are evolving into novel technologies for assessing drug metabolism, illness development, and docking molecular and chemical toxicology. Papers on metabolomics of medicinal plants have shown rapid expansion in the past few years, with a peak occurring in 2020; this has led to the broadening of the vision of plants, in terms of bioprospecting and biotechnology, and generating a complete perspective supported by metabolomic techniques as a multidisciplinary science.

Although metabolomics in medicinal plants is led by China, the USA, and Germany, a great global research potential is still found. In addition, the challenge in the coming years may be to strengthen the links between research and training centers that own technology and have access to high-quality analytical platforms, and with additional research centers in tropical countries to focus on the study of important and interesting local flora. New research must also be linked to promoting sustainability and conservation issues. As stated very clearly by the United Nations (UN), in the short to long term, advances in science, social progress, environmental balance, and economic growth must always progress hand in hand.

ACKNOWLEDGMENTS

AADVM acknowledges the financial support provided by SIP Instituto Politécnico Nacional Mexico (grant number IPN/SIP 20211305-20220810) and María asunción Bravo for technical support.

REFERENCES

Allard P. M., G. Genta-Jouve, and J. L. Wolfender. 2017. "Deep metabolome annotation in natural products research: Towards a virtuous cycle in metabolite identification." *Current Opinion in Chemical Biology* 36: 40–49. https://doi.org/10.1016/j.cbpa.2016.12.022

Arnald A., S. Marsal, and A. Julià. 2015. "Analytical methods in untargeted metabolomics: State of the art in 2015." *Frontiers in Bioengineering and Biotechnology* 3: 23. https://doi.org/10.3389/fbioe.2015.00023

Atanas G. A., B. Waltenberger, E. M. Pferschy-Wenzig, T. Linder, C. Wawrosch, P. Uhrin, V. Temml, L. Wang, S. Schwaiger, E. H. Heiss, J. M. Rollinger, D. Schuster, J. M. Breuss, V. Bochkov, M. D. Mihovilovic, B. Kopp, R. Bauer, V. M. Dirsch, and H. Stuppner. 2015. "Discovery and resupply of pharmacologically active plant-derived natural products: A review." *Biotechnology Advances* 33(8): 1582–1614. https://doi.org/10.1016/j.biotechadv.2015.08.001

Beale D. J., F. R. Pinu, K. A. Kouremenos, M. M. Poojary, V. K. Narayana, B. A. Boughton, K. Kanojia, S. Dayalan, O. A. H. Jones, and D. A. Dias. 2018. "Review of recent developments in GC–MS approaches to metabolomics-based research." *Metabolomics Official Journal of the Metabolomic Society* 14(11): 152. https://doi.org/10.1007/s11306-018-1449-2

Beger R. D., W. B. Dunn, A. Bandukwala, B. Bethan, D. Broadhurst, C. B. Clish, S. Dasari, L. Derr, A. Evans, S. Fischer, T. Flynn, T. Hartung, D. Herrington, R. Higashi, P. C. Hsu, C. Jones, M. Kachman, H. Karuso, G. Kruppa, K. Lippa, P. Maruvada, J. Mosley, I. Ntai, C. O'Donovan, M. Playdon, D. Raftery, D. Shaughnessy, A. Souza, T. Spaeder, B. Spalholz, F. Tayyari, B. Ubhi, M. Verma, T. Walk, I. Wilson, K. Witkin, D. W. Bearden and K. A. Zanetti. 2019. "Towards quality assurance and quality control in untargeted metabolomics studies." *Metabolomics Official Journal of the Metabolomic Society* 15(1): 4. https://doi.org/10.1007/s11306-018-1460-7

Benarba B., and A. Pandiella. 2020. "Medicinal plants as sources of active molecules against COVID-19." *Frontiers in Pharmacology* 11: 1189. https://doi.org/10.3389/fphar.2020.01189

Bingol K. 2018. "Recent advances in targeted and untargeted metabolomics by NMR and MS/NMR methods." *High-Throughput* 7(2): 9. https://doi.org/10.3390/ht7020009

Cai Y., K. Weng, Y. Guo, J. Peng, and Z. J. Zhu. 2015. "An integrated targeted metabolomic platform for high-throughput metabolite profiling and automated data processing." *Metabolomics* 11(6): 1575–1586. https://doi.org/10.1007/s11306-015-0809-4

Cai Z., H. Chen, J. Chen, R. Yang, L. Zou, C. Wang, J. Chen, M. Tan, Y. Mei, L. Wei, S. Yin, and X. Liu. 2020. "Metabolomics characterizes the metabolic changes of *Lonicerae japonicae* flos under different salt stresses." *PLOS ONE* 15(12): e0243111. https://doi.org/10.1371/journal.pone.0243111

Califf R. M. 2018. "Biomarker definitions and their applications." *Experimental Biology and Medicine* 243(3): 213–221. https://doi.org/10.1177/1535370217750088

Caspi R., R. Billington, C. A. Fulcher, I. M. Keseler, A. Kothari, M. Krummenacker, M. Latendresse, P. E. Midford, Q. Ong, W. K. Ong, S. Paley, P. Subhraveti, and P. D. Karp. 2018. "The MetaCyc database of metabolic pathways and enzymes." *Nucleic Acids Research* 46(D1): D633–D639. https://doi.org/10.1093/nar/gkx935.

Chen L., Y. Gao, L. Z. Wang, N. Cheung, G. S. W. Tan, G. C. M. Cheung, R. W. Beuerman, T. Y. Wong, E. C. Y. Chan, and L. Zhou. 2018. "Recent advances in the applications of metabolomics in eye research." *Analytica Chimica Acta* 1037: 28–40. https://doi.org/10.1016/j.aca.2018.01.060

Chikhale R. V., S. S. Gurav, R. B. Patil, S. K. Sinha, S. K. Prasad, A. Shakya, S. K. Shrivastava, N. S. Gurav, and R. S. Prasad. 2020. "Sars-cov-2 host entry and replication inhibitors from indian ginseng: An *in-silico* approach." *Journal of Biomolecular Structure and Dynamics*, 1–12. https://doi.org/10.1080/07391102.2020.1778539

Chong, Jasmine & Xia, Jianguo (Jeff). (2018). MetaboAnalystR: an R package for flexible and reproducible analysis of metabolomics data. *Bioinformatics* (Oxford, England). 34. 10.1093/bioinformatics/bty528.

Chua L. S. 2016. "Untargeted MS-based small metabolite identification from the plant leaves and stems of *Impatiens balsamina*." *Plant Physiology and Biochemistry* 106: 16–22. https://doi.org/10.1016/j.plaphy.2016.04.040

Cifuentes A. 2017. "Foodomics, foodome and modern food analysis." *Trends in Analytical Chemistry* 96: 1. https://doi.org/10.1016/j.trac.2017.09.001

Clish C. B. 2015. "Metabolomics: An emerging but powerful tool for precision medicine." *Cold Spring Harbor Molecular Case Studies* 1(1): 1–6. https://doi.org/10.1101/mcs.a000588

Cottret L., C. Frainay, M. Chazalviel, F. Cabanettes, Y. Gloaguen, E. Camenen, B. Merlet, S. Heux, J. C. Portais, N. Poupin, F. Vinson, and F. Jourdan. 2018. "MetExplore: Collaborative edition and exploration of metabolic networks." *Nucleic Acids Research* 46(W1): W495–W502. https://doi.org/10.1093/nar/gky301

Dehelean C. A., I. Marcovici, C. Soica, M. Mioc, D. Coricovac, S. Iurciuc, O. M. Cretu, and I. Pinzaru. 2021. "Plant-derived anticancer compounds as new perspectives in drug discovery and alternative therapy." *Molecules* 26(4): 1109. https://doi.org/10.3390/molecules26041109

Dong X., J. Fu, X. Yin, S. Cao, X. Li, L. Lin, and J. Ni. 2016. "Emodin: A review of its pharmacology, toxicity and pharmacokinetics." *Phytotherapy Research* 30(8): 1207–1218. https://doi.org/10.1002/ptr.5631

Emwas A. H., R. Roy, R. T. Mckay, L. Tenori, E. Saccenti, G. A. N. Gowda, D. Raftery, F. Alahmari, L. Jaremko, M. Jaremko, and D. S. Wishart. 2019. "NMR spectroscopy for metabolomics research." *Metabolites* 9(7): 123. https://doi.org/10.3390/metabo9070123

Fenaille F., P. B. Saint-Hilaire, K. Rousseau, and C. Junot. 2017. "Data acquisition workflows in liquid chromatography coupled to high resolution mass spectrometry-based metabolomics: Where do we stand?" *Journal of Chromatography* 1526: 1–12. https://doi.org/10.1016/j.chroma.2017.10.043

Fraser V. N., B. Philmus, and M. Megraw. 2020. "Metabolomics analysis reveals both plant variety and choice of hormone treatment modulate vinca alkaloid production in *Catharanthus roseus*." *Plant Direct* 4(9): 1–14. https://doi.org/10.1002/pld3.267

Gao D., M. Niu, S. Z. Wei, C. E. Zhang, Y. F. Zhou, Z. W. Yang, L. Li, J. B. Wang, H. Z. Zhang, L. Zhang, and X. H. Xiao. 2020. "Identification of a pharmacological biomarker for the bioassay-based quality control of a thirteen-component TCM formula (*Lianhua qingwen*) used in treating influenza a virus (H1N1) infection." *Frontiers in Pharmacology* 11: 746. https://doi.org/10.3389/fphar.2020.00746

Ghanbari J., G. Khajoei-Nejad, S. W. Erasmus, and S. M. van Ruth. 2019. "Identification and characterisation of volatile fingerprints of saffron stigmas and petals using PTR-TOF-MS: Influence of nutritional treatments and corm provenance." *Industrial Crops and Products* 141: 111803. https://doi.org/10.1016/j.indcrop.2019.111803

Ghassemi S., N. Delangiz, B. A. Lajayer, D. Saghafi, and F. Maggi. 2021. "Review and future prospects on the mechanisms related to cold stress resistance and tolerance in medicinal plants." *Acta Ecologica Sinica* 41(2): 120–129. https://doi.org/10.1016/j.chnaes.2020.09.006

Gonçalves J., T. Rosado, S. Soares, A. Simão, D. Caramelo, A. Luís, N. Fernández, M. Barroso, E. Gallardo, and A. Duarte. 2019. "*Cannabis* and its secondary metabolites: Their use as therapeutic drugs, toxicological aspects, and analytical determination." *Medicines* 6(1): 31. https://doi.org/10.3390/medicines6010031

Gonulalan E. M., E. Nemutlu, O. Bayazeid, E. Koçak, F. N. Yalçın, and L. O. Demirezer. 2020. "Metabolomics and proteomics profiles of some medicinal plants and correlation with BDNF activity." *Phytomedicine* 74. https://doi.org/10.1016/j.phymed.2019.152920

Gorrochategui E., J. Jaumot, S. Lacorte, and R. Tauler. 2016. "Data analysis strategies for targeted and untargeted LC-MS metabolomic studies: Overview and workflow." *Trends in Analytical Chemistry* 82: 425–442. https://doi.org/10.1016/j.trac.2016.07.004

Guijas C., J. R. Montenegro-Burke, X. Domingo-Almenara, A. Palermo, B. Warth, G. Hermann, G. Koellensperger, T. Huan, W. Uritboonthai, A. E. Aisporna, D. W. Wolan, M. E. Spilker, H. P. Benton, and G. Siuzdak. 2018. "METLIN: A technology platform for identifying knowns and unknowns." *Analytical Chemistry* 90(5): 3156–3164. https://doi.org/10.1021/acs.analchem.7b04424

Guo Y., R. Cao, X. Zhang, L. Huang, L. Sun, J. Zhao, J. Ma, and C. Han. 2019. "Recent progress in rare oncogenic drivers and targeted therapy for non-small cell lung cancer." *OncoTargets and Therapy* 12: 10343–10360. https://doi.org/10.2147/OTT.S230309

Hao D. C., and P. G. Xiao. 2015. "Genomics and evolution in traditional medicinal plants: Road to a healthier life." *Evolutionary Bioinformatics* 11: 197–212. https://doi.org/10.4137/EBO.S31326

Hasanpour M., M. Iranshahy, and M. Iranshahi. 2020. "The application of metabolomics in investigating anti-diabetic activity of medicinal plants." *Biomedicine and Pharmacotherapy* 128: 110263. https://doi.org/10.1016/j.biopha.2020.110263

Hellal K., A. Mediani, I. S. Ismail, C. P. Tan, and F. Abas. 2021. "^{1}H NMR-based metabolomics and UHPLC-ESI-MS/MS for the investigation of bioactive compounds from *Lupinus albus* fractions." *Food Research International* 140: 110046. https://doi.org/10.1016/j.foodres.2020.110046

Hellal K., M. Maulidiani, I. S. Ismail, C. P. Tan, and F. Abas. 2020. "Antioxidant, α-glucosidase, and nitric oxide inhibitory activities of six algerian traditional medicinal plant extracts and ^{1}H-NMR-based metabolomics study of the active extract." *Molecules* 25(5): 1247. https://doi.org/10.3390/molecules25051247

Hernandez-Perez T., M. R. Gómez-García, M. E. Valverde, and O. Paredes-Lopez. 2020. "*Capsicum annuum*: An ancient Latin-American crop with outstanding bioactive compounds and nutraceutical potential. A review." *Comprehensive Reviews in Food Science and Food Safety* 19: 2972–2993. https://doi.org/101111/154-4337.12634

Heyman H. M., and I. A. Dubery. 2016. "The potential of mass spectrometry imaging in plant metabolomics: A review." *Phytochemistry Reviews* 15(2): 297–316. https://doi.org/10.1007/s11101-015-9416-2

Holländer A. 2017. "Why do we need chemical derivatization?" *Plasma Processes and Polymers* 14(7): 1700044. https://doi.org/10.1002/ppap.201700044

Hussin M., A. A. Hamid, F. Abas, N. S. Ramli, A. H. Jaafar, S. Roowi, N. A. Majid and M. S. P. Dek. 2019. "NMR-based metabolomics profiling for radical scavenging and anti-aging properties of selected herbs." *Molecules* 24(17): 3208. https://doi.org/10.3390/molecules24173208

Jamwal K., S. Bhattacharya, and S. Puri. 2018. "Plant growth regulator mediated consequences of secondary metabolites in medicinal plants." *Journal of Applied Research on Medicinal and Aromatic Plants* 9: 26–38. https://doi.org/10.1016/j.jarmap.2017.12.003

Javadi N., F. Abas, A. Mediani, A. A. Hamid, A. Khatib, S. Simoh, and K. Shaari. 2015. "Effect of storage time on metabolite profile and alpha-glucosidase inhibitory activity of *Cosmos caudatus* leaves - GCMS based metabolomics approach." *Journal of Food and Drug Analysis* 23(3): 433–441. https://doi.org/10.1016/j.jfda.2015.01.005

Jia W., C. Wang, Y. Wang, G. Pan, M. Jiang, Z. Li, and Y. Zhu. 2015. "Qualitative and quantitative analysis of the major constituents in Chinese medical preparation Lianhua-Qingwen capsule by UPLC-DAD-QTOF-MS." *The Scientific World Journal* 2015: 731765. https://doi.org/10.1155/2015/731765

Kanehisa M., M. Furumichi, M. Tanabe, Y. Sato, and K. Morishima. 2017. "KEGG: New perspectives on genomes, pathways, diseases and drugs." *Nucleic Acids Research* 45(D1): D353–D361. https://doi.org/10.1093/nar/gkw1092

Karp P. D., R. Billington, R. Caspi, C. A. Fulcher, M. Latendresse, A. Kothari, I. M. Keseler, M. Krummenacker, P. E. Midford, Q. Ong, W. K. Ong, S. M. Paley, and P. Subhravet. 2019. "The BioCyc collection of microbial genomes and metabolic pathways." *Briefings in Bioinformatics* 20(4): 1085–1093. https://doi.org/10.1093/bib/bbx085

Keppler E. A. H., C. L. Jenkins, T. J. Davis, and H. D. Bean. 2018. "Advances in the application of comprehensive two-dimensional gas chromatography in metabolomics." *Trends in Analytical Chemistry* 109: 275–286. https://doi.org/10.1016/j.trac.2018.10.015

Khonakdari M. R, M. H. Mirjalili, A. Gholipour, H. Rezadoost, and M. M. Farimani. 2018. "Quantification of galantamine in *Narcissus tazetta* and *Galanthus vivalis* (Amaryllidaceae) populations growing wild in Iran." *Plant Genetic Resources: Characterization and Utilization* 16(2): 188–192. https://doi.org/10.1017/S1479262117000107

Khoo, L. W. A. Mediani, N. K. Z. Zolkeflee, S. W. Leong, I. S. Ismail, A. Khatib, K. Shaari, and F. Abas. 2015. "Phytochemical diversity of *Clinacanthus nutans* extracts and their bioactivity correlations elucidated by NMR based metabolomics." *Phytochemistry Letters* 14: 123–133. https://doi.org/10.1016/j.phytol.2015.09.015

Levin N., R. M. Salek, and C. Steinbeck. 2016. "From databases to big data." In *Metabolic Phenotyping in Personalized and Public Healthcare*, edited by E. Holmes, J. K. Nicholson, A. W. Darzi, and J. C. Lindon, 317–331. Elsevier. https://doi.org/10.1016/B978-0-12-800344-2.00011-2

Leyva-Padrón G, P. E. Vanegas-Espinoza, S. Evangelista-Lozano, A. A. Del Villar-Martínez, and C. Bazaldúa. 2020. "Chemical analysis of callus extracts from toxic and non-toxic varieties of *Jatropha curcas* L." *Peer J* 8: e10172. https://doi.org/10.7717/peerj.10172

Li A., J. Ma, H. Wang, F. Li, D. Qin, J. Wu, G. Zhu, J. Zhang, Y. Yuan, L. Zhou, and X. Wu. 2018. "NMR-based global metabolomics approach to decipher the metabolic effects of three plant growth regulators on strawberry maturation." *Food Chemistry* 269: 559–566. https://doi.org/10.1016/j.foodchem.2018.07.061

Li B. o., Tang Jing,Yang Qingxia, Li Shuang, Xuejiao Cui, Yinghong Li, Yuzong Chen, Weiwei Xue, Xiaofeng Li, Feng Zhu, 2017. NOREVA: normalization and evaluation of MS-based metabolomics data, *Nucleic Acids Research*, 45(W1), 3 July, Pages W162–W170, https://doi.org/10.1093/nar/gkx449

Liang Q., C. Wang, B. Li, and A. Zhang. 2015. "Metabolic fingerprinting to understand therapeutic effects and mechanisms of silybin on acute liver damage in rat." *Pharmacognosy Magazine* 11(43): 586–593. https://doi.org/10.4103/0973-1296.160469

Ma X., Y. Meng, P. Wang, Z. Tang, H. Wang, and T. Xie. 2020. "Bioinformatics-assisted, integrated omics studies on medicinal plants." *Briefings in Bioinformatics* 21(6): 1857–1874. https://doi.org/10.1093/bib/bbz132

Matveeva, T., G. Khafizova, and S. Sokornova. 2020. "In search of herbal anti-SARS-Cov2 compounds." *Frontiers in Plant Science* 11: 589998. https://doi.org/10.3389/fpls.2020.589998

Mir S. A., G. Jeyabalan, G. Parveen, and M. U. Rahman. 2020. "Hepatoprotective activity of *Mentha arvensis* in anti-tubercular drugs induced hepatotoxicity in rats." *Research in Pharmacy and Health Sciences* 6(2): 115–120. https://doi.org/10.32463/RPHS.2020.v06i02.01

Moldoveanu S. C., and V. David. 2019. "Derivatization methods in GC and GC/MS." In *Gas Chromatography - Derivatization, Sample Preparation, Application*, edited by P. Kusch. IntechOpen. https://doi.org/10.5772/intechopen.81954

Mumtaz M. W., A. A. Hamid, M. T. Akhtar, F. Anwar, U. Rashid, and M. H. AL-Zuaidy. 2017. "An overview of recent developments in metabolomics and proteomics - Phytotherapic research perspectives." *Frontiers in Life Science* 10(1): 1–37. https://doi.org/10.1080/21553769.2017.1279573

Murugesu S., Z. Ibrahim, Q. U. Ahmed, B. F. Uzir, N. I. N. Yusoff, V. Perumal, F. Abas, K. Shaari, and A. Khatib. 2019. "Identification of α-glucosidase inhibitors from *Clinacanthus nutans* leaf extract using liquid chromatography-mass spectrometry-based metabolomics and protein-ligand interaction with molecular docking." *Journal of Pharmaceutical Analysis* 9(2): 91–99. https://doi.org/10.1016/j.jpha.2018.11.001

Naik P. M., and J. M. Al-Khayri. 2016. "Abiotic and biotic elicitors - Role in secondary metabolites production through *in vitro* culture of medicinal plants." *Abiotic and Biotic Stress in Plants - Recent Advances and Future Perspectives*: 247–277. https://doi.org/10.5772/61442

Nemadodzi L. E., J. Vervoort, and G. Prinsloo. 2020. "NMR-based metabolomic analysis and microbial composition of soil supporting *Burkea africana* growth." *Metabolites* 10(10): 1–17. https://doi.org/10.3390/metabo10100402

Okada T., Afendi F. M. , Altaf-Ul-Amin M., Takahashi H., Nakamura K., Kanaya S. (2010) Metabolomics of medicinal plants: the importance of multivariate analysis of analytical chemistry data. *Curr Comput Aided Drug Des* Sep; 6(3):179–96. doi: 10.2174/157340910791760055. PMID: 20550511.

Patten G. S., M. Y. Abeywardena, and L. E. Bennett. 2016. "Inhibition of angiotensin converting enzyme, angiotensin II receptor blocking, and blood pressure lowering bioactivity across plant families." *Critical Reviews in Food Science and Nutrition* 56(2): 181–214. https://doi.org/10.1080/10408398.2011.651176

Pérez-Mendoza M. B., L. Llorens-Escobar, P. E. Vanegas-Espinoza, A. Cifuentes, E. Ibáñez, and A. A. Del Villar-Martínez. 2019. "Chemical characterization of leaves and calli extracts of *Rosmarinus officinalis* by UHPLC-MS." *Electrophoresis* 41(20): 1776–1783. https://doi.org/10.1002/elps.201900152

Perianes-Rodriguez A., L. Waltman, and N. J. van Eck. 2016. "Constructing bibliometric networks: A comparison between full and fractional counting." *Journal of Informetrics* 10(4): 1178–1195. https://doi.org/10.1016/j.joi.2016.10.006

Persson M. S. M., J. Stocks, D. A. Walsh, M. Doherty, and W. Zhang. 2018. "The relative efficacy of topical non-steroidal anti-inflammatory drugs and capsaicin in osteoarthritis: A network meta-analysis of randomised controlled trials." *Osteoarthritis and Cartilage* 26(12): 1575–1582. https://doi.org/10.1016/j.joca.2018.08.008

Qamar M. T. U., S. M. Alqahtani, M. A. Alamri, and L. L. Chen. 2020. "Structural basis of SARS-CoV-2 3CL^{pro} and anti-COVID-19 drug discovery from medicinal plants." *Journal of Pharmaceutical Analysis* 10(4): 313–319. https://doi.org/10.1016/j.jpha.2020.03.009

Rashid A., V. Ali, M. Khajuria, S. Faiz, S. Gairola, and D. Vyas. 2021. "GC-.MS based metabolomic approach to understand nutraceutical potential of *Cannabis seeds* from two different environments." *Food Chemistry* 339: 128076. https://doi.org/10.1016/j.foodchem.2020.128076

Rastogi S., S. Shah, R. Kumar, A. Kumar, and A. K. Shasany. 2020. "Comparative temporal metabolomics studies to investigate interspecies variation in three *Ocimum* species." *Scientific Reports* 10: 5234. https://doi.org/10.1038/s41598-020-61957-5

Rathahao-Paris E., S. Alves, C. Junot, and J. C. Tabet. 2016. "High resolution mass spectrometry for structural identification of metabolites in metabolomics." *Metabolomics* 12: 10. https://doi.org/10.1007/s11306-015-0882-8

Reddy P. R. K., M. M. M. Y. Elghandour, A. Z. M. Salem, D. Yasaswini, P. P. R. Reddy, A. N. Reddy, and I. Hyder. 2020. "Plant secondary metabolites as feed additives in calves for antimicrobial stewardship." *Animal Feed Science and Technology* 264: 114469. https://doi.org/10.1016/j.anifeedsci.2020.114469

Reel P. S., S. Reel, E. Pearson, E. Trucco, and E. Jefferson. 2021. "Using machine learning approaches for multi-omics data analysis: A review." *Biotechnology Advances* 49: 107739. https://doi.org/10.1016/j.biotechadv.2021.107739

Reyes-Vaquero L., M. Bueno, R. I. Ventura-Aguilar, A. B. Aguilar-Guadarrama, N. Robledo, G. Sepúlveda-Jiménez, P. E. Vanegas-Espinoza, E. Ibáñez, and A. A. Del Villar-Martínez. 2021. "Seasonal variation of chemical profile of *Ruta graveolens* extracts and biological activity against *Fusarium oxysporum*, *Fusarium proliferatum* and *Stemphylium vesicarium*." *Biochemical Systematics and Ecology* 95: 104223. https://doi.org/10.1016/j.bse.2021.104223

Rochat B. 2016. "From targeted quantification to untargeted metabolomics: Why LC-high-resolution-MS will become a key instrument in clinical labs." *Trends in Analytical Chemistry* 84: 151–164. https://doi.org/10.1016/j.trac.2016.02.009

Runfeng L., H. Yunlong, H. Jicheng, P. Weiqi, M. Qinhai, S. Yongxia, L. Chufang, Z. Jin, J. Zhenhua, J. Haiming, Z. Kui, H. Shuxiang, D. Jun, L. Xiaobo, H. Xiaotao, W. Lin, Z. Nanshan, and Y. Zifeng. 2020. "Lianhuaqingwen exerts anti-viral and anti-inflammatory activity against novel coronavirus (SARS-CoV-2)." *Pharmacological Research* 156: 104761. https://doi.org/10.1016/j.phrs.2020.104761

Salem M. A., R. A. Radwan, E. S. Mostafa, S. Alseekh, A. R. Fernie, and S. M. Ezzat. 2020. "Using an UPLC/MS-based untargeted metabolomics approach for assessing the antioxidant capacity and anti-aging potential of selected herbs." *RSC Advances* 10(52): 31511–31524. https://doi.org/10.1039/D0RA06047J

Salmerón-Manzano E., J. A. Garrido-Cardenas, and F. Manzano-Agugliaro. 2020. "Worldwide research trends on medicinal plants." *International*

Journal of Environmental Research and Public Health 17(10): 3376. https://doi.org/10.3390/ijerph17103376
Samani M. R., G. D'Urso, P. Montoro, A. G. Pirbalouti, and S. Piacente. 2021. "Effects of bio-fertilizers on the production of specialized metabolites in *Salvia officinalis* L. leaves: An analytical approach based on LC-ESI/LTQ-Orbitrap/MS and multivariate data analysis." *Journal of Pharmaceutical and Biomedical Analysis* 197: 113951. https://doi.org/10.1016/j.jpba.2021.113951
Seca A. M. L., and D. C. G. A. Pinto. 2018. "Plant secondary metabolites as anticancer agents: Successes in clinical trials and therapeutic application." *International Journal of Molecular Sciences* 19(1): 263. https://doi.org/10.3390/ijms19010263
Seenivasan A., J. S. Eswari, P. Sankar, S. N. Gummadi, T. Panda, and C. Venkateswarlu. 2020. "Metabolic pathway analysis and dynamic macroscopic model development for lovastatin production by *Monascus purpureus* using metabolic footprinting concept." *Biochemical Engineering Journal* 154: 107437. https://doi.org/10.1016/j.bej.2019.107437
Sharifi, N., Souri, E., Ziai, S. A., Amin, G., and Amanlou, M. (2013). Discovery of new angiotensin converting enzyme (ACE) inhibitors from medicinal plants to treat hypertension using an in vitro assay. *Daru* 21 (1), 74. https://doi.org/10.1186/2008-2231-21-74
Sharma Y., R. Velamuri, J. Fagan, and J. Schaefer. 2020. "UHPLC-ESI-QTOF-Mass spectrometric assessment of the polyphenolic content of *Salvia officinalis* to evaluate the efficiency of traditional herbal extraction procedures." *Revista Brasileira de Farmacognosia* 30(5): 701–708. https://doi.org/10.1007/s43450-020-00106-5
Sherriff S. C., J. S. Rowan, O. Fenton, P. Jordan, and D. ÓhUallacháin. 2018. "Sediment fingerprinting as a tool to identify temporal and spatial variability of sediment sources and transport pathways in agricultural catchments." *Agriculture, Ecosystems and Environment* 267: 188–200. https://doi.org/10.1016/j.agee.2018.08.023
Srinivasan, K. 2016. "Biological activities of red pepper (*Capsicum annuum*) and its pungent principle capsaicin: A review." *Critical Reviews in Food Science and Nutrition* 56(9): 1488–1500. https://doi.org/10.1080/10408398.2013.772090.
Steuer A. E., L. Brockbals, and T. Kraemer. 2019. "Metabolomic strategies in biomarker research-new approach for indirect identification of drug consumption and sample manipulation in clinical and forensic toxicology?" *Frontiers in Chemistry* 7: 319. https://doi.org/10.3389/fchem.2019.00319
Stilo F., C. Bicchi, A. M. Jimenez-Carvelo, L. Cuadros-Rodriguez, S. E. Reichenbach, and C. Cordero. 2021. "Chromatographic fingerprinting by comprehensive two-dimensional chromatography: Fundamentals and tools." *TrAC Trends in Analytical Chemistry* 134: 116133. https://doi.org/10.1016/j.trac.2020.116133
Tenenboim H., and Y. Brotman. 2016. "Omic relief for the biotically stressed: Metabolomics of plant biotic interactions." *Trends in Plant Science* 21(9): 781–791. https://doi.org/10.1016/j.tplants.2016.04.009

Thakur M., S. Bhattacharya, P. K. Khosla, and S. Puri. 2019. "Improving production of plant secondary metabolites through biotic and abiotic elicitation." *Journal of Applied Research on Medicinal and Aromatic Plants* 12: 1–12. https://doi.org/10.1016/j.jarmap.2018.11.004

Tripathi M. K., P. Singh, S. Sharma, T. P. Singh, A. S. Ethayathulla, and P. Kaur. 2020. "Identification of bioactive molecule from *Withania somnifera* (Ashwagandha) as SARS-CoV-2 main protease inhibitor." *Journal of Biomolecular Structure and Dynamics* 39(15): 5668–5681. https://doi.org/10.1080/07391102.2020.1790425

Tsakadze D. M., A. Abdusamatov, and S. Yu. Yunusov. 1969. "Alkaloids of *Galanthus aucasicus*." *Chemistry of Natural Compounds* 5: 281–282. https://doi.org/10.1007/BF00683870

Urquiza-López A., G. Álvarez-Rivera, D. Ballesteros-Vivas, A. Cifuentes, and A. A. Del Villar-Martínez. 2021. "Metabolite profiling of rosemary cell lines with antiproliferative potential against human HT-29 colon cancer cells." *Plant Foods for Human Nutrition* 76: 319–325. https://doi.org/10.1007/s11130-021-00892-w

Viant M. R, I. J. Kurland, M. R. Jones, and W. B. Dunn. 2017. "How close are we to complete annotation of metabolomes?" *Current Opinion in Chemical Biology* 36: 64–69. https://doi.org/10.1016/j.cbpa.2017.01.001

Wanichthanarak K., S. Fan, D. Grapov, D. K. Barupal, and O. Fiehn. 2017. "Metabox: A toolbox for metabolomic data analysis, interpretation and integrative exploration." *PLoS One* 12(1): e0171046. https://doi.org/10.1371/journal.pone.0171046

Wang W., J. Xu, H. Fang, Z. Li, and M. Li. 2020. "Advances and challenges in medicinal plant breeding." *Plant Science* 298: 110573. https://doi.org/10.1016/j.plantsci.2020.110573

Wang X. J., J. L. Ren, A. H. Zhang, H. Sun, G. Yan, Y. Han, and L. Liu. 2019. "Novel applications of mass spectrometry-based metabolomics in herbal medicines and its active ingredients: Current evidence." *Mass Spectrometry Reviews* 38(4–5): 380–402. https://doi.org/10.1002/mas.21589

Yan M., and G. Xu. 2018. "Current and future perspectives of functional metabolomics in disease studies – A review." *Analytica Chimica Acta* 1037: 41–54. https://doi.org/10.1016/j.aca.2018.04.006

Yang J., Q. Lin, L. Fan, and N. Yang. 2020. "High performance liquid chromatography-quadrupole time-of-flight mass spectrometry based metabolomic detection of non-volatile components of different chemotype of *Cinnamomum camphora*." *Journal of Analytical Chemistry* 75(12): 1582–1588. https://doi.org/10.1134/S1061934820120138

Yu M. S., J. Lee, J. M. Lee, Y. Kim, Y. W. Chin, J. G. Jee, Y. S. Keum, and Y. J. Jeong. 2012. "Identification of myricetin and scutellarein as novel chemical inhibitors of the SARS coronavirus helicase, NsP13." *Bioorganic and Medicinal Chemistry Letters* 22(12): 4049–4054. https://doi.org/10.1016/j.bmcl.2012.04.081

Yuan L., and E. Grotewold. 2020. "Plant specialized metabolism." *Plant Science* 298: 110579. https://doi.org/10.1016/j.plantsci.2020.110579

Zhao Q., J. L. Zhang, and F. Li. 2018. "Application of metabolomics in the study of natural products." *Natural Products and Bioprospecting* 8(4): 321–334. https://doi.org/10.1007/s13659-018-0175-9

Zhou G., and J. Xia. 2018. "OmicsNet: A web-based tool for creation and visual analysis of biological networks in 3D space." *Nucleic Acids Research* 46(W1): W514–W522. https://doi.org/10.1093/nar/gky510

Zolkeflee N. K. Z., N. A. Isamail, M. Maulidiani, N. A. A. Hamid, N. S. Ramli, A. Azlan, and F. Abas. 2021. "Metabolite variations and antioxidant activity of *Muntingia calabura* leaves in response to different drying methods and ethanol ratios elucidated by NMR-based metabolomics." *Phytochemical Analysis* 32(1): 69–83. https://doi.org/10.1002/pca.2917

CHAPTER 6

Extraction, Encapsulation, and Biological Activity of Phenolic Compounds, Alkaloids, and Acetogenins from the *Annona* Genus

Yolanda Nolasco-González,
Luis Miguel Anaya-Esparza,
Gabriela Aguilar-Hernández,
Brandon Alexis López-Romero, and
Efigenia Montalvo-González

CONTENTS

DOI: 10.1201/9781003166535-6

6.1 INTRODUCTION

The Annonaceae family members are angiosperms and are distributed mainly in the tropical and subtropical regions of Central and South America, Africa, Asia, and Australia. The height of the trees is 3–8 m with reddish and hairless trunks, branches, and stem barks. Lenticels on the trunk are numerous, cylindrical, and rough. Leaves are oblong-lanceolate, dark green, and 5–15 cm long. Flowers are paired, axillary or supra-axillary, appearing on trunks or branches, are unisexual, often with a single adaxial bract. Fruits reach maturity in 12–17 weeks, depending on *Annona* species, with an average weight from 100 g to 2.9 kg. Peels are thin, green to dark green, and covered with soft spines. Pulp is white, relatively fibrous, and very aromatic. Seeds (170–250 per fruit) are obovoid, flattened, and 15–20 mm long with dark testa (Rodríguez-Nuñez et al. 2021). *A. muricata* (soursop), *A. cherimola*, and *A. squamosa* plants are native to Mesoamerica and West Indies. They are the fruits which are mainly grown (Coria-Téllez et al. 2017; Anaya-Esparza et al. 2020; Rodríguez-Nuñez et al. 2021). The consumption and commercialization of *Annona* fruits have attracted attention due to their exotic taste and flavor. Moreover, infusion or decoction of leaves, stem bark, and roots of these species have been used for different ailments (Coria-Téllez et al. 2018; Anaya-Esparza et al. 2020). Therefore, diverse studies have been focused on the extraction and encapsulation of bioactive compounds from these plants, and to investigate their biological activity (Jordán-Suárez et al. 2018; Aguilar-Hernández et al. 2019, 2020a).

Studies on the extraction of bioactive compounds from the three species have increased in recent years (Aguilar-Hernández et al. 2020b) due to their biological activities, such as antioxidant, anti-tumoral, anti-inflammatory, and anti-hyperglycemic activities, among others (Coria-Téllez et al. 2018; Anaya-Esparza et al. 2020). Some extraction technologies, namely maceration, heat reflux, and ultrasound, have been proposed to obtain extracts. Moreover, different encapsulation methods have been used to protect the chemical structures of bioactive compounds, guaranteeing health-related benefits (Aguilar-Hernández et al. 2019, 2020a).

This chapter aimed to summarize the current scientific advances on phenolic compounds, alkaloids, and acetogenins from *A. muricata*, *A. cherimola*, and *A. squamosa*, taking into account the main extraction and encapsulation technologies, the *in vitro* and *in vivo* biological activities, as well as the potential applications.

6.2 TRADITIONAL USES OF *ANNONA* GENUS

The *Annona* genus comprises >70 species, of which those being studied in this chapter are highlighted as the species which are commercially cultivated for their fruits worldwide. They are used for human consumption and as medicines (Coria-Tellez et al. 2017; Anaya-Esparza et al. 2020; Rodríguez-Nuñez et al. 2021). In this context, the *Annona* fruits have a subacid taste and have juicy, sweet, creamy, and fragrant pulps, which are typically consumed fresh or frozen. Furthermore, *Annona* fruits are used as functional ingredients to elaborate beverages and food products such as nectars, juices, purées, yogurts, milkshakes, and ice cream, among others (Anaya-Esparza et al. 2020).

Additionally, *Annona* species are used as decoctions or infusions from leaves, stem bark, and roots due to their medicinal properties. Extracts have been shown to treat diverse illnesses, such as fever, cold, flu, asthma, diarrhea, urethritis, and dysentery, among others (Coria-Tellèz et al. 2018). The traditional uses are associated with the biological activities of the already known secondary metabolites such as phenolic compounds, alkaloids, and acetogenins (Anaya-Esparza et al. 2020).

6.3 PHENOLIC COMPOUNDS, ALKALOIDS, AND ACETOGENINS FROM *ANNONA* SPECIES

Phenolic compounds are biosynthesized by the shikimic acid and "polyketide" acetate pathways. The general chemical structure is represented by one or more phenolic rings (Figure 6.1). Forty-five types of

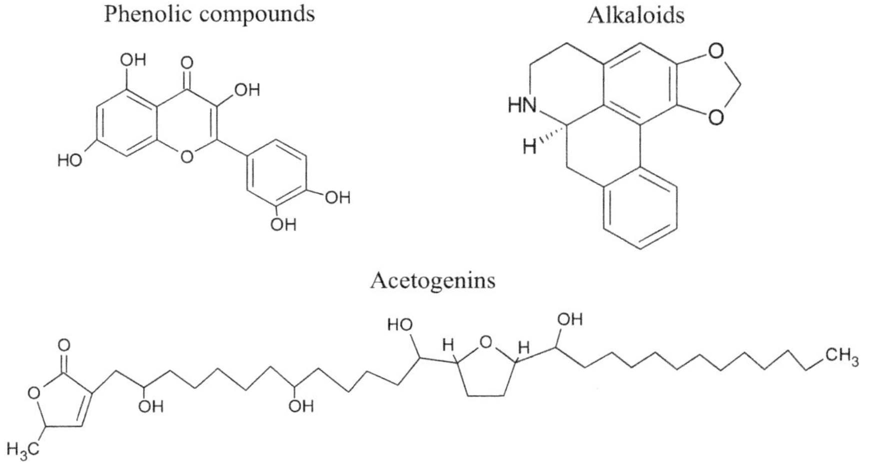

FIGURE 6.1 General chemical structures of representative phenolic compounds (Pubchem ID: 5280343), alkaloids (Pubchem ID: 160597), and acetogenins (Pubchem ID: 354398) identified in *A. muricata*, *A. cherimola*, and A. *squamosa* (Figures created with Chem Sketch v.2).

phenolic compounds have been reported in *A. muricata*, *A. cherimola*, and *A. squamosa*, the main functions of which in plants are as antioxidants, antimicrobial agents, and pollinator attractants (Aguilar-Hernández et al. 2019; Mannino et al. 2020; Shehata et al. 2021). On the other hand, alkaloids are heterocyclic nitrogenous compounds derived mainly from lysine, tyrosine, tryptophan, and ornithine (Figure 6.1). In the reviewed *Annona* species, more than 27 reported benzylisoquinoline alkaloids have been reported, which are of type aporphine, protoberberines, benzyloquinoleins, and bisbenzylisoquinolein dimers. Meanwhile, acetogenins are secondary metabolites exclusive to the Annonaceae, which are biosynthesized by the polyketide pathway. Acetogenins are molecules of a long aliphatic chain of 35 to 37 carbon atoms with one, two, or three tetrahydrofuran or tetrahydropyran rings in the central region (Figure 6.1). Annonacin and bullatacin are present at the highest concentrations of more than 120 acetogenins identified. These secondary metabolites have been isolated from different organs and tissues of adult plants, and are used as insecticides and antimicrobial agents (Pinto et al. 2017; Galarce-Bustos et al. 2020; Lee et al. 2021; Nugraha et al. 2021). In addition, they are important from a pharmaceutical point of view because several beneficial activities have been demonstrated for different illnesses.

6.4 EXTRACTION AND ENCAPSULATION METHODS OF PHENOLIC COMPOUNDS, ALKALOIDS, AND ACETOGENINS FROM *ANNONA* SPECIES

Figure 6.2 shows that, from by-products, fruits, leaves, stem bark, and roots of *A. muricata*, *A. cherimola*, and *A. squamosa*, phenolic compounds, alkaloids, and acetogenins have been extracted using maceration, the Soxhlet method, or ultrasound-assisted extraction. Furthermore, various efforts have been made to protect these secondary metabolites from environmental factors, using encapsulation methods to increase bioactivity, bioavailability, and effectiveness (Aguilar-Hernández et al. 2019; 2020a, b; Aguilar-Villalva et al. 2021).

The yield of secondary metabolites achieved by each extraction method depends on thermostability, solvent polarity, mass transfer, and raw material. In this context, maceration and Soxhlet methods are conventional techniques to extract bioactive compounds. However, ultrasound-assisted extraction is considered to be an emerging, viable, and eco-friendly

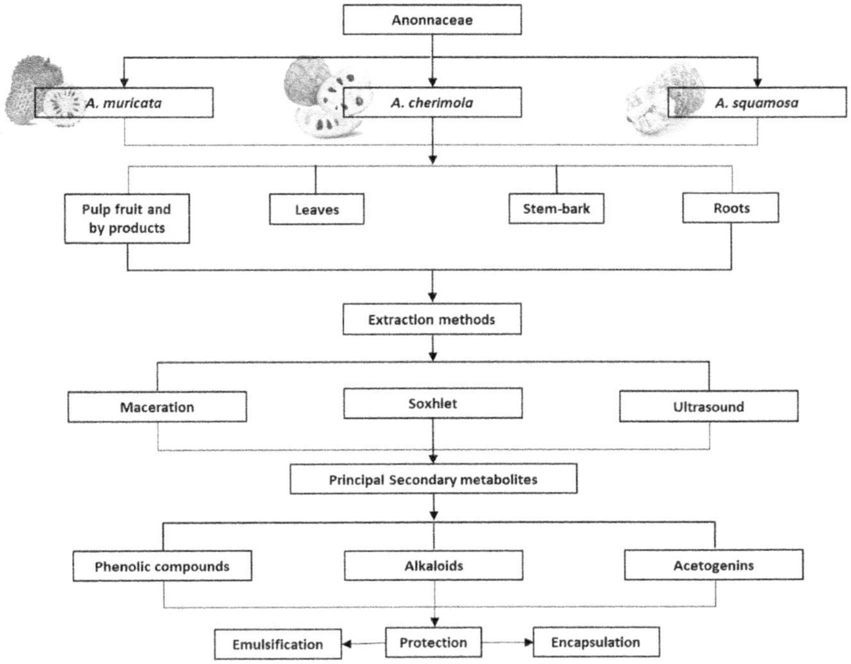

FIGURE 6.2 Identification, extraction, and protection of bioactive compounds from different tree parts, fruits, and by-products of some *Annona* species.

technology to extract bioactive compounds (Aguilar-Hernández et al. 2019; Aguilar-Villalva et al. 2021).

Maceration is a conventional and inexpensive method that uses organic solvents and long extraction times to recover bioactive compounds from plant materials (Ibrahim et al. 2020). This extraction method has been used to extract phenolic compounds, alkaloids, and acetogenins from fruits, seeds, peels, stem bark, and leaves of *A. muricata*, *A. cherimola*, and *A. squamosa* (Benarba et al. 2018; Ibrahim et al. 2020). Alkaloids, flavonoids, saponins, and tannins from *A. muricata* leaf powder have been extracted by aqueous, ethanolic, and methanolic maceration for 24 h (Syed-Najmuddin et al. 2017; Ibrahim et al. 2020), whereas acetogenins have been extracted using methanolic, hexanic, chloroformic, and ethyl acetate solutions for seven days (Ismail et al. 2018).

The Soxhlet method is widely used to extract analytes from solid samples. This method has been used to obtain crude extracts (hydrophilic and hydrophobic) from *Annona* leaves, stem bark, roots, fruits, and by-products. The solvents utilized to obtain polyphenols or alkaloids from *Annona* species are water, ethanol, or methanol, while acetogenins are extracted with ethanol, methanol, hexane, dichloromethane, ethyl acetate, or chloroform (Chen et al. 2016; Veerakumar et al. 2016). However, the maceration and Soxhlet methods main limitations are the large volumes of organic solvents used, long extraction times, and low yields obtained (León-Fernández et al. 2017).

Ultrasound-assisted extraction is an emerging method based on acoustic cavitation, and is widely used to obtain bioactive compounds from plant materials in shorter times than are needed with conventional methods (Aguilar-Villalva et al. 2021). Aguilar-Hernández et al. (2019) reported that ultrasound-assisted extraction was a technological alternative for obtaining phenolic compounds from *A. muricata* peel using water-methanolic solutions during 15–30 min extractions, achieving a higher yield (37.51%) than that obtained under conventional aqueous-organic extraction (16.68%). Moreover, similar trends were reported in the alkaloid yields from *A. muricata* peel using methanol and acetogenins from *A. muricata* seeds using chloroform, obtaining 56.31 times more alkaloids and 9.93 times more acetogenins using ultrasound than with maceration (Aguilar-Hernández et al. 2020a, b).

Additionally, ultrasound-assisted extraction has been used to extract polyphenols from *A. cherimola* leaves, significantly reducing the extraction time to 15 min compared to 48 h for the Soxhlet method and seven days for maceration (Aguilar-Villalva et al. 2021). Furthermore, ultrasound-assisted extraction has been employed to obtain the highest yield of polyphenols (2.68%) from *A. squamosa* peel in comparison with the Soxhlet method (1.96%) and maceration (2.07%) (Deng et al. 2015). In this context, ultrasound-assisted extraction is, therefore, a viable technology to extract bioactive compounds from *Annona* peel, seeds, and leaves, although the yield depends on the raw materials.

As mentioned in previous sections, phenolic compounds, alkaloids, and acetogenins extracted from *A. muricata*, *A. cherimola*, and *A. squamosa* are traditionally used in ethnomedicine. However, polyphenols and alkaloids could be degraded after extraction because they are sensitive to light, oxygen, and high temperatures. Therefore, some strategies, such as encapsulation through emulsification and the use of other carriers in the form of nanoparticles, have been studied to preserve the chemical composition, solubility, bioavailability, and further pharmaceutical applications of these bioactives. Encapsulation is one of the most common methods for preserving bioactive compounds. This technique covers bioactive compounds with a homogeneous or heterogeneous polymeric matrix to create macro, micro, or nanocapsules (Gutiérrez et al. 2020). Jordán-Suárez et al. (2018) prepared spherical capsules loaded with an extract from *A. muricata* leaves by the spray-drying method. The authors used maltodextrin as an encapsulating agent. These capsules exhibited up to 98% solubility of bioactive compounds in a concentration-dependent manner and with suitable characteristics for pharmaceutical applications. Recently, it has been reported that encapsulation by spray-drying is a viable alternative to preserve the antioxidant properties of an extract from *A. muricata* leaves (Jordán-Suárez et al. 2021) and fresh juice pulp (Neta et al. 2019).

A. muricata fruit pulp was encapsulated in calcium alginate hydrogels. Micrographs obtained by scanning electron microscopy showed that the incorporation of 20% soursop pulp preserved the capsule morphology. The authors concluded that calcium alginate capsules loaded with soursop pulp could be useful as healthy food ingredients (López-Córdoba et al. 2016). Similarly, Cornelia et al. (2019) encapsulated polyphenols extracted from *A. muricata* leaves, using *Aloe vera* mucilage as an encapsulating agent. The authors demonstrated the highest encapsulation efficiency under these conditions (88–91%) and concluded that capsules prolong the antioxidant activity of biomolecules. Furthermore, freeze-dried chitosan matrices are an alternative for the slow release of acetogenins obtained from *A. muricata* during *in vitro* gastrointestinal digestion (Mulia et al. 2019).

Another technological strategy is the incorporation of bioactive compounds in lipidic phases via emulsification to preserve stability (Abd-Elghany and Mohamad 2020). Mancini et al. (2018) developed phytosomes (emulsions of phosphatidylcholine and plant extracts) functionalized with phenolic compounds extracted from *A. muricata* leaves. Phytosomes protected the phenolic compounds during passage through the digestive tract, preventing biotransformation and increasing the permeability capacity of the blood-brain barrier. Furthermore, by electroporation, Abd-Elghany and Mohamad (2020) formulated an *A. squamosa* leaf extract entrapped in niosomes (vesicle carriers formed by nonionic surfactants). The bioavailability of *A. squamosa* compounds significantly increased because loaded niosomes passed through the dermal/transdermal skin. The increase in permeability started from 15 min to 45 min after skin electroporation. Gutiérrez et al. (2020) developed supramolecular polymer micelles that were loaded with annonacin (SPMA) isolated from *A. cherimola* leaves. They had an encapsulation efficiency of 35%. Micelles increased the cytotoxic activity of annonacin (100 μM, 24 h, 94.2% of dead cells) compared with non-encapsulated annonacin (100 μM, 24 h, 92% of living cells).

In general, these technologies are an alternative to preserve, enhance the administration, and application of isolated compounds or raw extracts; however, more investigations on encapsulation methods or carriers of bioactive compounds from *Annona* species are necessary due to limited research reports.

6.5 BIOLOGICAL ACTIVITIES FROM EXTRACTS OF *A. MURICATA, A. CHERIMOLA,* AND *A. SQUAMOSA*

This section discusses the bioactive compounds and the biological activities reported from *Annona* species (*A. muricata*, *A. cherimola*, and *A. squamosa*) mainly associated with the presence of polyphenols, alkaloids, and acetogenins.

6.5.1 Antioxidant Capacities of Extracts of *A. muricata*, *A. cherimola*, and *A. squamosa*

Antioxidant capacity is the capability of a substance to inhibit oxidative degradation (lipid peroxidation, for instance). Antioxidants, called chain terminator antioxidants, act with free radicals (George et al. 2015). Therefore, the interest in obtaining natural antioxidant compounds from *Annona* species has increased in recent years, associated with protective effects against various non-communicable diseases (Coria-Tellez et al.

2018). Diverse studies have demonstrated that aqueous, ethanolic, methanolic, and acetonic extracts from leaves, seeds, pulp, and stem bark of *A. muricata*, *A. cherimola*, and *A. squamosa* exhibited antioxidant capacity. According to these reports, the *in vitro* antioxidant capacity is dependent on *Annona* species, doses, type, and concentration of phenolic compounds and other antioxidants present, such as cyclopeptides, terpenes, and ascorbic acid content (Ojezele et al. 2016; Nguyen et al. 2020; Aguilar-Villalva et al. 2021).

In vitro, the antioxidant capacity of *Annona* extracts or pulps indicates that, when humans consume them, they act as antioxidants suitable for use as exogenous antioxidant supplements to increase detoxification mechanisms. Therefore, *Annona* antioxidants are another alternative to prevent or counteract a high level of oxidative stress, mainly in people that suffer from chronic degenerative diseases (Shehata et al. 2021).

6.5.2 Cytotoxic Activity from Extracts or Isolated Acetogenins of *A. muricata, A. cherimola,* and *A. squamosa*

Cytotoxic activity has been demonstrated from *A. muricata, A. cherimola,* and *A. squamosa* extracts or isolated acetogenins (Table 6.1). It has been reported that different types of acetogenins or extracts obtained from *A. muricata* exhibited potent cytotoxic effects against various cancer cell lines (liver, breast, colon, lung, prostate, and leukemia) in a dose-dependent response (Coria-Tellez et al. 2018).

The cytotoxic effect of annomuricin E or *n*-butanolic extracts obtained from *A. muricata* leaves was observed in lung and prostate cancer cell lines (George et al. 2015; Zorofchian-Moghadamtousi et al. 2015). Annonacin from *A. muricata* leaves exhibited a potent cytotoxic effect through the *in vitro* MTT [3-(4, 5-dimethylthiazol-2-yl)-2,5-diphenyltetrazolium bromide] assay, reporting a half-maximal inhibitory concentration (IC_{50}) with a value of 2.9 ± 1.3 μM (Roduan et al. 2019).

Hadisaputri et al. (2021) evaluated the cytotoxic potential of *A. muricata* leaf extracts (with ethanol, ethyl acetate, *n*-hexane, or water solvents) on a cell line of human breast adenocarcinoma line MCF-7; extracts showed high cytotoxic activity with IC_{50} values of 5.3, 2.86, 3.08, and 48.31 μg/mL, respectively. Furthermore, the authors reported changes in the morphology of the cancer cells 6 h after application of the extracts. It is inferred that the cytotoxic activity is due to the decrease in the abundance of *Bcl-2* mRNA and an increase in the abundance of caspase-9 and caspase-3 mRNAs, causing cell rupture and loss of nuclei.

TABLE 6.1 Cytotoxic Activity of Different Extracts or Isolated Compounds of *A. muricata*, *A. cherimola*, or *A. squamosa*

Annona Species	Plant Part	Extract/Bioactive Compound	Dose/Concentration	Model Assay	Effect	References
A. muricata	Leaf	Annomuricin E	IC_{50} 1.62 µg/mL	Colon cancer cell line implanted in rats	Apoptosis	Zorofchian-Moghadamtousi et al. 2015
	Leaf	n-Butanolic extract	Not reported	Human breast carcinoma	Potent cytotoxic effects	George et al. 2015
	Leaf	Annonacin	IC_{50} 2.9 ± 1.3 µM	*In vitro* MTT assay	Selective cytotoxic effects	Roduan et al. 2019
	Pulp	Solid lipidic nanoparticles with ethanolic extracts	IC_{50} 12.1 g/mL	*In vitro* MTT assay on cell lines (MCF-7)	Apoptosis	Sabapati et al. 2019
	Leaf	Silver nanoparticles with aqueous *Annona* extracts	IC_{50} 57.37 µg/ml	*In vitro* MTT assay on human keratinocyte cell lines (HaCaT)	Apoptosis	Badmus et al. 2020
	Leaf	Ethanol, ethyl acetate, *n*-hexane, and water extracts	IC_{50} 5.3, 2.86, 3.08, and 48.31 µg/mL, respectively	*In vitro* MTT assay on cell lines (MCF-7).	Apoptosis	Hadisaputri et al. (2021)
A. cherimola	Stem bark	Annonacin and ethyl acetate	IC_{50} 0.1 µM and 55.501 µg/mL, respectively	Prostate carcinoma cells	Cytotoxic effects	Foster et al. 2020
	Seeds	Ethanolic	IC_{50} ranged from 50 to 100 µg/mL	Acute Myeloid Leukemia (AML) cell lines *in vitro*, KG-1, Monomac-1 and U937	Apoptosis	Haykal et al. 2019
	Seeds	Ethanolic	23.2 and 30.9 µg/mL	*In vitro* with colorectal cancer (CRC) cell lines T84, HCT-15,	Selective cytotoxic effects	Fuel et al. 2021

(*Continued*)

TABLE 6.1 (CONTINUED) Cytotoxic Activity of Different Extracts or Isolated Compounds of *A. muricata*, *A. cherimola*, or *A. squamosa*

Annona Species	Plant Part	Extract/Bioactive Compound	Dose/Concentration	Model Assay	Effect	References
A. squamosa	Leaf	Ethanolic	400 µg/mL	Human colon cancer cell line	Potent cytotoxicity	Fadholly et al. 2019
	Leaf Seeds	Methanolic	IC_{50} ranged from 1.1 to 2.1 µg/mL	Breast cancer cell line	Potent cytotoxicity	Pinto et al. 2017
	Seeds Pulp Peel	Aqueous	IC_{50} ranged from 7.31 to 15.99 µg/mL	Colon, prostate, liver, and breast cancer cell lines	Selective cytotoxic effects	Shehata et al. 2021

On the other hand, silver nanoparticles with extracts of *A. muricata* leaves have been formulated. These nanoparticles presented different biological activities, one of these being cytotoxic activity, which was assayed in the human keratinocyte cell line HaCaT. The positive controls used were tamoxifen and cisplatin for 24 h. The IC_{50} cytotoxicity value was 57.37 μg/mL for nanoparticles, 3.12 μg/mL for cisplatin, and 22.00 μg/mL of tamoxifen. The cytotoxic effect was attributed to the active interaction between phytoconstituents and silver atoms with the functional groups of intracellular proteins, nitrogenous bases, and phosphate groups of the DNA. In addition, the cytotoxic effect can also be ascribed to cellular damage, induction of a cascade of immunological effects, and electrostatic attraction between cells and nanoparticles (Badmus et al. 2020).

Similar results were found in solid lipid nanoparticles using ethanolic extracts from *A. muricata* pulp, with lecithin as a surfactant, and stereoacetic acid as a lipid matrix. The cytotoxicity was determined in the MCF7 cell line (human breast adenocarcinoma). Higher cytotoxic activity of nanoparticles loaded with extracts (IC_{50} 12.1 g/mL) was observed than with the free extract (IC_{50} 30.1 g/mL). This increase may be due to the high permeability of nanoparticles into the cellular matrix of cancer cells, which is related to the nanometric size of encapsulates. In addition to the composition of these nanoparticles, it has been reported that the lecithin used as a surfactant contributes to improving the permeability characteristics. Based on these results, lipid nanoparticles loaded with bioactive compounds from *A. muricata* extracts have been proposed as an alternative to oral and intravenous administration (Sabapati et al. 2019).

Fuel et al. (2021) reported that ethanolic extracts from *A. cherimola* seeds showed a selective cytotoxic effect against the colorectal cancer cell lines, T84 and HCT-15, with the IC_{50} values being 23.2 μg/mL and 30.9 μg/mL, respectively. Likewise, proapoptotic results have also been reported in acute myeloid leukemia cell lines KG-1, Monomac-1, and U937, revealing IC_{50} values of 57, 107, and 100 μg/mL, respectively (Haykal et al. 2019). Recently, Foster et al. (2020) reported that ethyl acetate extract or annonacin from *A. cherimola* stem bark showed selective and potent cytotoxicity against prostate carcinoma cells.

In addition, ethanolic and methanolic extracts of *A. squamosa* leaves and seeds showed potent cytotoxic effects against both colon and breast cancer cell lines (Pinto et al. 2017; Fadholly et al. 2019). Similar trends were reported in colon, prostate, liver, and breast cancer cell lines after being treated with aqueous extracts from seeds, pulp, and peel of *A.*

squamosa, and the effect was due to a dose-dependent response (Shehata et al. 2021).

In general, acetogenins could be potential alternatives for treating cancer. These compounds reduce ATP production by inhibiting the mitochondrial nicotinamide adenine dinucleotide ubiquinone oxidase reductase (NADH oxidase reductase) of complex I of the respiratory chain, and promoting apoptosis (Coria-Tellez et al. 2018).

6.5.3 Anti-Inflammatory Activity from Extracts of *A. muricata*, *A. cherimola*, and *A. squamosa*

The anti-inflammatory activity of diverse extracts from *A. muricata*, *A. cherimola*, and *A. squamosa* has been extensively investigated (Table 6.2).

Oyekachukwu et al. (2017) investigated the anti-inflammatory effect of a chloroform extract of *A. muricata* leaves on the *in vitro* activity of phospholipase A_2 and prostaglandin synthase, as well as on membrane stabilization. These authors reported an inhibition of between 23.9 and 43.5% of the enzyme activity of phospholipase A_2 when 0.2–0.6 mg/mL of chloroform extract was used. A dose of 0.5 mg/mL showed 87.5% of inhibition of prostaglandin synthase activity compared with indomethacin (1.0 mg/mL), which is used as an anti-inflammatory. On the other hand, the extract at 0.4 mg/mL inhibited the heat-induced hemolysis of human red blood cells by 53.0%, with the highest inhibition of hypotonicity-induced hemolysis (77.9%) occurring at a concentration of 0.8 mg/mL. This study shows that the mechanism of action of the chloroform extract is due to the high concentration of phytoconstituents such as soluble phenolics and flavonoids, acting via inhibition of the synthesis of inflammatory mediators. Phenolic compounds can inhibit the enzymes by binding to active sites; the hydroxyl groups (–OH) or acylated sugar could bind to the polar charges of the erythrocyte membranes with consequent protection, avoiding degradation (Ruiz-Ruiz et al. 2017). The inhibition of phospholipase A_2 activity suppressed the release of free fatty acids from the membrane with the consequent loss of cyclooxygenase and lipoxygenase precursors for the synthesis of inflammatory mediators, limiting effects such as vasodilatation, vascular permeability, chemotaxis, and pain, thus preventing inflammation. Therefore, the active agent might serve as an adjuvant for conventional drugs.

According to Bitar et al. (2019), the aqueous extract of *A. muricata* leaves at 100 mg/kg of body weight may reduce the inflammation process by decreasing pro-inflammatory cytokine expression promoted by

TABLE 6.2 Anti-inflammatory and Anti-hyperglycemic Activities of Different Crude Extracts or Isolated Compounds of *A. muricata*, *A. cherimola*, and *A. squamosa*

Activity/*Annona* Species	Plant part	Extract/ Bioactive Compound	Dose/ Concentration	Model Assay	Effect	References
Anti-inflammatory						
	Leaf	Chloroform	0.6 mg/mL for phospholipase and 0.5 mg/ mL for prostaglandin synthase	*In vitro* phospholipase A_2 and prostaglandin synthase	The extract inhibited phospholipase A2 and prostaglandin synthase activity by 43.5% and 87.46%, respectively	Oyekachukwu et al. 2017
A. muricata	Leaf	Aqueous	100 mg/kg body weight	Rats challenged with *Escherichia fergusonii*	The extract reduced the inflammation process by decreasing cytokine expression	Bitar et al. 2019
	Leaf	Ethanolic	100, 200, 400 mg/kg body weight	*In vivo* anti-inflammation against hemorrhoids in mice induced by croton oil	Significant anti-inflammatory effects on hemorrhoidal tissue, decreased COX-2 and TNF-α.	Ayun et al. 2020
	Leaf	Ethanolic	1000 μg/mL	*In vitro* inhibition of albumin denaturation	Comparable results to commercial drugs in a dose-dependent response	Amaud et al. 2020
A. cherimola	Leaf	Ethanolic	100 mg/kg body weight	Rats (leukocyte migration to the peritoneal cavity and subcutaneous air pouch test)	Significant anti-inflammatory effect in a dose-dependent response	do-Nascimento-Silva et al. 2017

(*Continued*)

TABLE 6.2 (CONTINUED) Anti-inflammatory and Anti-hyperglycemic Activities of Different Crude Extracts or Isolated Compounds of *A. muricata*, *A. cherimola*, and *A. squamosa*

Activity/*Annona* Species	Plant part	Extract/ Bioactive Compound	Dose/ Concentration	Model Assay	Effect	References
	Seed	Cherimolacy-clopeptide B analogs	20–40 µg/mL	*In vitro* Inhibitory effects on the production of pro-inflammatory cytokines using lipopolysaccharide-stimulated macrophage J774A.1 cell line	Cherimolacyclopedtide analogs had anti-inflammatory activity suppressing the secretion of Interleukin IL-6 and Tumor necrosis factor (TNF-α)	Dahiya and Dahiya 2021
A. squamosa	Leaf	Aqueous	300 mg/kg body weight for four weeks	Rats (acetic acid-induced colitis)	The extract counteracted ulcerative colitis by decreasing malondialdehyde (MDA) concentration in colonic tissue and increasing catalase (CAT), glutathione (GSH), and glutathione peroxidase (Gpx).	Ibrahim et al. 2015
	Leaf	Ethanolic, Aqueous	300 mg/kg body weight	Rats (aluminum chloride-induced neuroinflammation)	Extracts reduced the brain inflammation	Hendawy et al. 2019
A. squamosa	Peel	Ethanolic	400 mg/kg body weight/ day	Rats (models of rheumatoid of arthritis)	The extract exhibited anti-inflammatory properties in a time-dependent response	Phan et al. 2021

(Continued)

TABLE 6.2 (CONTINUED) Anti-inflammatory and Anti-hyperglycemic Activities of Different Crude Extracts or Isolated Compounds of *A. muricata*, *A. cherimola*, and *A. squamosa*

Activity/*Annona* Species	Plant part	Extract/ Bioactive Compound	Dose/ Concentration	Model Assay	Effect	References
Anti-hyperglycemic						
A. muricata	Pulp	Fresh pulp	100 g of pulp	*In vitro* assay	Pulp exhibited potential hypoglycemic properties	Passos et al. 2015
	Leaf	Aqueous	10 μL of extract of the extract obtained at 100°C for 30 min	*In vitro* anti-diabetic activity by α-glucosidase enzyme inhibition	The extract had an IC_{50} 396.70 ppm for the highest α-glucosidase enzyme inhibition	Hardoko et al. 2015
	Leaf	Ethanol	25 to 80 μg/mL	*In vitro* enzymatic assay using α-amylase	Extract significantly inhibited the α-amylase activity	Justino et al. 2018
A. cherimola	Leaf	Ethanolic/ Rutin	300 mg/kg body weight	Rats (alloxan-induced diabetes type-2)	Extracts reduced blood glucose levels	Calzada et al. 2017
	Leaf	Infusion	300 mg/kg body weight	Rats (alloxan-induced diabetes type-2)	Extracts reduced blood glucose levels	Vasarri et al. 2020
	Leaf	Infusion	300 mg/kg body weight	Healthy and diabetic subjects	Pulp consumption did not promote postprandial glycemia	Martínez-Solís et al. 2021

(*Continued*)

TABLE 6.2 (CONTINUED) Anti-inflammatory and Anti-hyperglycemic Activities of Different Crude Extracts or Isolated Compounds of *A. muricata*, *A. cherimola*, and *A. squamosa*

Activity/*Annona* Species	Plant part	Extract/ Bioactive Compound	Dose/ Concentration	Model Assay	Effect	References
A. squamosa	Peel	Ethanolic	10–100 μg/mL	*In vitro* α-amylase inhibitory test, healthy rats	The extract exhibited hepatoprotective effects	Roy and Sasikala 2016
	Peel Leaf	Aqueous	250 mg/kg body weight	Rats (Streptozotocin- (STZ) induced diabetic)	Extracts decreased glycemia and improved insulin tolerance	Sahu et al. 2016
	Leaves	Ethanol, aqueous	300 mg/kg body weight	Albino Wistar male rats (alloxan-induced diabetes)	The mix of extracts had the highest reduction of fasting blood sugar (FBS) and increased the insulin level	El-Baz, Khayat and Hssam 2019

Escherichia fergusonii in a challenged rat model. On the other hand, a study using the albumin denaturation inhibition method demonstrated that the anti-inflammatory effect of an *A. muricata* ethanolic extract was similar to those obtained using commercial anti-inflammatory drugs like diclofenac sodium (Amaud et al. 2020). Nonetheless, extracts from *A. muricata* leaves improve venous tone and hemorrhoidal anti-inflammation in a dose-dependent response. Ayun and Elya (2020) reported the anti-inflammatory effect of the ethanolic extract of *A. muricata* leaves against hemorrhoids in mice induced by croton oil. The ethanolic extract, dosed orally for seven days at doses of 100, 200, or 400 mg/kg body weight, resulted in different levels of inflammation, necrosis, and vasodilation in the histopathology of rectoanal mice tissue. Also, 200 and 400 mg/kg doses significantly decreased the cyclooxygenase (COX-2) activity. In contrast, all doses decreased the concentration of the tumor necrosis factor TNF-α, indicating that the extract can reduce inflammation and potentially be a natural remedy for hemorrhoids.

Additionally, do Nascimento-Silva et al. (2017) reported that the ethanolic extract from leaves of "Atemoya", a hybrid between *A. cherimola* and *A. squamosa*, at a dose of 100 mg/kg body weight, exhibited a significant anti-inflammatory effect on carrageenan-induced peritonitis on rats. In the anti-inflammatory model, the extract caused 63.85% inhibition of the migration of leukocytes in the peritoneal cavity. Dahiya and Dahiya (2021) reported the inhibitory effect of a cherimolacyclopeptide B analog extracted from *A. cherimola* seeds on the production of pro-inflammatory cytokines induced by lipopolysaccharide (LPS) in the macrophage J774A.1 cell line. The substitution of methionine and glycine of cherimolacyclopeptide B with alanine resulted in novel peptide analogs that showed significant anti-inflammatory activity by suppressing the secretion of interleukin (IL-6) and tumor necrosis factor (TNF-α). Ibrahim et al. (2015) evaluated the anti-inflammatory activity of an aqueous extract from *A. squamosa* leaves against rats exhibiting acetic acid-induced colitis. Results showed that a dose of 300 mg/kg body weight for 4 weeks counteracted the induced ulcerative colitis by decreasing malondialdehyde accumulation and increasing the activities of antioxidant enzymes catalase, glutathione, and glutathione peroxidase in the colonic tissue. Thus, the *A. squamosa* extract can be considered to be an agent for alleviating colitis. Phan et al. (2021) reported that oral administration of an ethanolic extract of *A. squamosa* peel at 400 mg/kg body weight/day exhibited anti-inflammatory activity in rheumatoid arthritis-induced rats. Moreover, ethanolic and aqueous extracts from *A. squamosa* leaves administered at 300 mg/kg body weight reduced brain inflammation in aluminum chloride-induced neuroinflammation rats (Hendawy et al. 2019).

According to the data, the anti-inflammatory property of *Annona* extracts is attributed to diverse bioactive compounds in extracts, including phenolic compounds, kaurenes or sesquiterpenes, and alkaloids, which can inhibit inflammatory mediators like cyclooxygenases (COX-1 and COX-2), lipoxygenases, and central nociceptors associated with opioid properties, inhibiting the serotonin and histamine release (Oyekachukwu et al. 2017).

6.5.4 Anti-Hyperglycemic Activity from Extracts of *A. muricata, A. cherimola,* and *A. squamosa*

The *in vitro* and *in vivo* anti-hyperglycemic activities of leaves, roots, peel, and pulp extracts from *A. muricata*, *A. cherimola*, and *A. squamosa* have also been investigated, although most of the reported studies are focused on extracts from leaves (Table 6.2). Diverse *in vitro* studies have demonstrated that the anti diabetic activity of aqueous extracts of *A. muricata* leaves is achieved by inhibiting some enzymes such as α-glucosidase (Hardoko et al. 2015) and α-amylase using ethanolic extract (Justino et al. 2018). The α-glucosidase enzyme plays a role in starch and glycogen metabolism, so that reducing the activity of this enzyme could decrease the hydrolysis of carbohydrates. In this context, Hardoko et al. (2015) elaborated an infusion from *A. muricata* leaves (incubated at 100 °C for 30 min) that contained phenols (205.37 mg/L), tannins (100.33 mg/L), and flavonoids (99.97 mg/L). The infusion inhibited the α-glucosidase enzyme, achieving the lowest IC_{50} value (396 mg/L). The authors concluded that the infusion from *A. muricata* leaves has potential as an anti diabetic functional drink. On the other hand, Justino et al. (2018) obtained phenolic-enriched fractions (ethyl acetate and *n*-butanol) from an ethanolic extract of *A. muricata* leaves, which were used in enzymatic assays to evaluate their inhibitory potential against α-amylase, α-glucosidase, and lipase. The ethyl acetate and *n*-butanol fractions were a rich source of chlorogenic and caffeic acids, procyanidins B2 and C1, epicatechin, quercetin, quercetin-hexoside, and kaempferol. The ethyl acetate and *n*-butanol fractions showed inhibitory capacities of α-amylase (IC_{50} 9.2 and 6.1 μg/mL, respectively), α-glucosidase (IC_{50} 413.1 and 817.4 μg /mL, respectively) and lipase (IC_{50} 74.2 and 120.3 μg/mL, respectively) (Justino et al. 2018).

A similar trend has been reported in alloxan-induced hyperglycemic rats after administration of a methanolic extract from *A. muricata* seeds at doses of 200 to 600 mg/kg body weight. Moreover, 100 g of soursop (*A. muricata*) pulp consumption reduced the glycemic index in healthy subjects from 150 to 90 mg/dL of glucose. This effect was attributed to the polyphenolic content (Passos et al. 2015).

In terms of *A. cherimola*, the ethanolic extract (Calzada et al. 2017) or aqueous infusion (Vasarri et al. 2020; Martínez-Solís et al. 2021) of leaves exhibited anti diabetic properties in streptozotocin-induced hyperglycemic rats at a dose of 300 mg/kg body weight. In general, the extract significantly reduced the blood glucose levels with improved function on glycosylated hemoglobin, cholesterol, triglycerides, and high-density lipoprotein content without modification of the urine profile after treatment of hyperglycemic rats (Martínez-Solís et al. 2021). Nonetheless, the consumption of 100 g/day of *A. cherimola* pulp did not contribute to an increase in postprandial glycemia in either healthy or diabetic subjects (Vasarri et al. 2020).

In addition, it has been reported that aqueous (Sahu et al. 2016) and methanolic (Kumar, Chandra and Gajera 2019) extracts of *A. squamosa* leaves improved the glucose tolerance and decreased glycemia, with an increase in insulin tolerance of streptozotocin-induced and alloxan-induced diabetic rats in a dose-dependent response from 100 to 500 mg/kg body weight. Also, El-Baz et al. (2019) reported the effect of ethanolic and aqueous extracts of *A. squamosa* leaves on alloxan-induced hyperglycemia rats. Results showed that 300 mg/kg body weight/day significantly decreased the concentration of fasting blood sugar, serum cholesterol, triglycerides, low-density lipoprotein, liver enzymes, serum urea, and creatinine, and the mixture of ethanol and aqueous extracts had a greater effect than individual extracts.

Aqueous extracts from *A. squamosa* peel and leaves were administrated at a dose of 250 mg/kg body weight to streptozotocin-induced hyperglycemic rats. The leaf extract caused significant improvements in plasma glucose, triglycerides, total cholesterol, glucose utilization, and insulin tolerance activity of blood compared with the peel extract (Sahu et al. 2016). Likewise, an *in vitro* study demonstrated that an ethanolic extract of *A. squamosa* peel exhibited an α-amylase inhibitory effect (Roy and Sasikala 2016). In this study, before ethanolic extraction, the peel was treated with steam blanching to evaluate secondary metabolites. There was a reduction in alkaloids, tannins, and saponins when the blanching time increased. In contrast, flavonoids increased in response to three to five minutes of blanching, and, subsequently, α-amylase inhibition decreased in these samples with IC_{50} values from 3.31 to 5.53 μg/mL (Roy and Sasikala 2016).

In general, the reduction of blood glucose levels is associated with phenolic compounds from *Annona* extracts that mitigate diabetes-induced oxidative stress (Justino et al. 2018). In this regard, phenolic compounds can activate antioxidant enzymatic systems, such as glutathione-S-transferase, catalase, glutathione reductase, superoxide dismutase, and non-enzymatic antioxidant systems, like glutathione (Kumar et al. 2017). Moreover, these extracts increased insulin secretion, modulated the digestion of the carbohydrates, and inhibited intestinal glucose absorption (Sahu et al. 2016). These results reinforced the traditional use of *Annona* extracts against diabetes mellitus.

6.6 CONCLUSION

Significant evidence indicates that *A. muricata*, *A. cherimola*, and *A. squamosa* contain biologically active compounds such as polyphenols, alkaloids, and acetogenins. Ultrasound-assisted extraction is a method that extracts the highest yield of these compounds and is a sustainable technology that can be scaled up to obtain *Annona* bioactive compounds at the industrial scale. Phenolic compounds, alkaloids, and acetogenins can be used in diverse pharmaceutical applications due to their antioxidant, cytotoxic, anti-inflammatory, and anti-hyperglycemic properties. The reported information supports the widespread importance of natural extracts of these *Annona* species in traditional medicine. Additionally, isolated bioactive compounds obtained from *Annona* species can be encapsulated to preserve stability and enhance their biological effects. In this context, the extraction, characterization, and preservation of bioactive compounds of different *Annona* species is an active research area. However, further studies are needed to understand the mechanism of action of the many secondary metabolites to guarantee safe use and increase their yield in industrial production.

ACKNOWLEDGMENTS

This review is part of the activities of the Annonaceae National Network.

Conflict of interest: Authors declare no conflict of interest.

REFERENCES

Abd-Elghany, A. A., and E. A. Mohamad. 2020. *Ex-vivo* transdermal delivery of *Annona squamosa* entrapped in niosomes by electroporation. *Journal of Radiation Research and Applied Sciences* 13, no. 1: 164–73. https://doi.org/10.1080/16878507.2020.1719329

Aguilar-Hernández, G., L. G. Zepeda-Vallejo, M. L. García-Magaña, M. A. Vivar-Vera, A. Pérez-Larios, M. I. Girón-Pérez, A. V Coria-Téllez, C. Rodríguez-Aguayo, and E. Montalvo-González. 2020a. Extraction of alkaloids using ultrasound from pulp and by-products of soursop fruit (*Annona muricata L.*). *Applied Sciences* 10, no. 14: 4869. https://doi.org/10.3390/app10144869

Aguilar-Hernández, G., M. A. Vivar-Vera, M. L. García-Magaña, N. González-Silva, A. Pérez-Larios, and E. Montalvo-González. 2020b. Ultrasound-assisted extraction of total acetogenins from the soursop fruit by response surface methodology. *Molecules* 25, no. 5: 1139. http://doi.org/10.3390/molecules25051139

Aguilar-Hernández, G., M. L. García-Magaña, M. A. Vivar-Vera, S. G. Sáyago-Ayerdi, J. A. Sánchez-Burgos, J. Morales-Castro, L. M. Anaya-Esparza, and E. Montalvo González. 2019. Optimization of ultrasound-assisted extraction of phenolic compounds from *Annona muricata* by-products and pulp. *Molecules* 24, no. 5: 904. https://doi.org/10.3390/molecules24050904

Aguilar-Villalva, R., G. A. Molina, L. F. Díaz-Peña, A. Elizalde-Mata, E. Valerio, C. Azanza-Ricardo, and M. Estevez. 2021. Antioxidant capacity and antibacterial activity from *Annona cherimola* phytochemicals by ultrasound-assisted extraction and its comparison to conventional methods. *Arabian Journal of Chemistry* 14, no. 1: 103239. https://doi.org/10.1016/j.arabjc.2021.103239

Amaud, K., C. Nicodème, D. Durand, N. Martial, S. Basile, S. Haziz, N. Christine, K. Christian, L. Halfane, D. Victorien, P. Noumavo, and B. Lamine. 2020. Antioxidant efficacy, anti-inflammatory and HPLC analysis of *Annona muricata* leaf extracts from the Republic of Benin. *American Journal of Plant Sciences* 11, no. 6: 803–18. https://www.scirp.org/journal/paperinformation.aspx?paperid=101082

Anaya-Esparza, L. M., M. L. García-Magaña, J. A. Domínguez-Ávila, E. M. Yahia, N. J. Salazar-López, G. A. González-Aguilar, and E. Montalvo-González. 2020. *Annonas*: underutilized species as a potential source of bioactive compounds. *Food Research International* 138: 109775. https://doi.org/10.1016/j.foodres.2020.109775

Ayun, N. Q., and B. Elya. 2020. Anti-inflammation of soursop leaves (*Annona muricata* L.) against hemorrhoids in mice induced by croton oil. *Pharmacognosy Journal* 12, no. 4: 784–92. https://www.phcogj.com/article/1176

Badmus, J. A., S. A. Oyemomi, O. T. Adedosu, T. A. Yekeen, M. A. Azeez, E. A. Adebayo, A. Lateef, U. M. Badeggi, S. Botha, A. A. Hussein, and J. L. Marnewic. 2020. Photo-assisted bio-fabrication of silver nanoparticles using *Annona muricata* leaf extract: Exploring the antioxidant, anti-diabetic, antimicrobial, and cytotoxic activities. *Heliyon* 6, no. 11: 1–9. https://doi.org/10.1016/j.heliyon.2020.e05413

Benarba, B., O. I. N. Mendas, and S. Righi. 2018. Phytochemical analysis, antioxidant and anti-*Candida albicans* activities of *Annona cherimola* Mill. fruit pulp. *The North African Journal of Food and Nutrition Research* 2, no. 4: 120–8. http://doi.org/10.5281/zenodo.1495218

Bitar, R. M., R. R. Fahmi, and J. M. Borjac. 2019. *Annona muricata* extract reduces inflammation via inactivation of NALP3 inflammasome. *Journal of Natural Remedies* 19, no. 1: 12–23. https://doi.org/10.18311/jnr/2018/22617

Calzada, F., J. I. Solares-Pascasio, R. M. Ordoñez-Razo, C. Velazquez, E. Barbosa, N. García-Hernández, D. Mendez-Luna, and J. Correa-Basurto. 2017. Antihyperglycemic activity of the leaves from *Annona cherimola* miller and rutin on alloxan-induced diabetic rats. *Pharmacognosy Research* 9, no. 1: 1–6. https://doi.org/10.4103/0974-8490.199781

Chen, Y., Y. Chen, Y. Shi, C. Ma, X. Wang, Y. Li, Y. Miao, J. Chen, and X. Li. 2016. Antitumor activity of *Annona squamosa* seed oil. *Journal of Ethnopharmacology* 193, no. 4: 362–7. https://doi.org/10.1016/j.jep.2016.08.036

Coria-Téllez, A. V., E. Montalvo-González, E. M. Yahia, and E. N. Obledo-Vázquez. 2018 *Annona muricata*: A comprehensive review on its traditional medicinal uses, phytochemicals, pharmacological activities, mechanisms of action and toxicity. *Arabian Journal of Chemistry* 11, no. 5: 662–91. https://doi.org/10.1016/j.arabjc.2016.01.004

Coria-Téllez, A. V., E. Montalvo-González, and E. N. Obledo-Vázquez. 2017. Soursop (*Annona muricata*). In *Fruit and Vegetable Phytochemicals: Chemistry and Human Health*, edited by E. M. Yahia, 1243–52. Oxford: Wiley Online Library. https://doi.org/10.1002/9781119158042.ch66

Cornelia, M., N. Kam, H. Cahyana, and E. Sutiyono. 2019. Encapsulation of soursop (*Annona muricata* Linn.) leaf tea extract using natural mucilage. *Reaktor* 19, no. 1: 26–33. https://doi.org/10.14710/reaktor.19.1.26-33

Dahiya, R., and S. Dahiya. 2021. Natural bioeffective cyclooligopeptides from plant seeds of *Annona* genus. *European Journal of Medicinal Chemistry* 214, no. 15: 113221. https://doi.org/10.1016/j.ejmech.2021.113221

Deng, G. F., D. P. Xu, S. Li, and H. B. Li. 2015. Optimization of ultrasound-assisted extraction of natural antioxidants from sugar apple (*Annona squamosa* L.) peel using response surface methodology. *Molecules* 20, no. 11: 20448–59. https://doi.org/10.3390/molecules201119708

do Nascimento-Silva, H., S. V. Rabêlo, T. C. Diniz, F. G. S. Oliveira, R. B. A. Teles, J. C. Silva, M., G. Silva, H. D. M. Coutinho, I. R. A. Menezes, and J. R. G. S. Almeida. 2017. Antinociceptive and anti-inflammatory activities of ethanolic extract from atemoya (*Annona cherimola* Mill x *Annona squamosa* L.). *African Journal of Pharmacy and Pharmacology* 11, no. 18: 224–32. https://doi.org/10.5897/AJPP2017.4778

El-Baz, D. M., Z. El-Khayat, and A. K. Hssan. 2019. Effects of Egyptian *Annona squamosa* leaves extracts against alloxan-induced hyperglycemia in rats. *World Journal of Pharmacy and Pharmaceutical Sciences* 8, no. 3: 145–63. https://doi.org/10.20959/wjpps20193-13297

Fadholly, A., A. Proboningrat, R. P. D. Iskandar, F. A. Rantam, and S. A. Sudjarwo. 2019. *In vitro* anti-cancer activity *Annona squamosa* extract nanoparticle on WiDr cells. *Journal of Advanced Pharmaceutical Technology and Research* 10, no. 4: 149–54. https://doi.org/10.4103/japtr.JAPTR_10_19

Foster, K., O. Oyenihi, S. Rademan, J. Erhabor, M. Matsabisa, J. Barker, M. K. Langat, A. K. Smith, H. Asemota, and R. Delgoda. 2020. Selective cytotoxic and anti-metastatic activity in DU-145 prostate cancer cells induced by *Annona muricata* L. bark extract and phytochemical, annonacin. *BMC Complementary Medicine and Therapies* 20, no. 1: 1–15. https://doi.org/10.1186/s12906-020-03130-z

Fuel, M., C. Mesas, R. Martínez, R. Ortiz, F. Quiñonero, J. Prados, J. Porres, and C. Melguizo. 2021. Antioxidant and antiproliferative potential of ethanolic extracts from *Moringa oleifera*, *Tropaeolum tuberosum* and *Annona cherimola* in colorrectal cancer cells. *Biomedicine and Pharmacotherapy* 143, no. 1: 753. https://doi.org/10.1016/j.biopha.2021.112248

Galarce-Bustos, O., M. T. Fernández-Ponce, A. Montes, C. Pereyra, L. Casas, C. Mantell, and M. Aranda. 2020. Usage of supercritical fluid techniques to obtain bioactive alkaloid-rich extracts from cherimoya peel and leaves: Extract profiles and their correlation with antioxidant properties and acetylcholinesterase and α-glucosidase inhibitory activities. *Food and Function* 11, no. 5: 4224–35. https://doi.org/10.1039/D0FO00342E

George, V. C., R. Kumar, P. K. Suresh, and R. A. Kumar. 2015. Antioxidant, DNA protective efficacy and HPLC analysis of *Annona muricata* (soursop) extracts. *Journal of Food Science and Technology* 52, no. 4: 2328–35. https://doi.org/10.1007/s13197-014-1289-7

Gutiérrez, M. T., A. G. Durán, F. J. Mejías, J. M. Molinillo, D. Megias, M. M. Valdivia, and F. A. Macías. 2020. Bio-guided isolation of acetogenins from *Annona cherimola* deciduous leaves: Production of nanocarriers to boost the bioavailability properties. *Molecules* 25, no. 20: 4861. https://doi.org/10.3390/molecules25204861

Hadisaputri, Y. E., U. Habibah, F. F. Abdullah, E. Halimah, M. Mutakin, S. Megantara, R. Abdulah, and A. Diantini. 2021. Antiproliferation activity and apoptotic mechanism of soursop (*Annona muricata* L.) leaves extract and fractions on MCF7. *Breast Cancer Cells* 13, no. 1: 447–57. https://doi.org/10.2147/BCTT.S317682

Hardoko, Y. H., S. V. Wijoyo, and Y. Halim. 2015. *In vitro* anti-diabetic activity of "green tea" soursop leaves brew through α-glucosidase inhibition. *International Journal of PharmTech Research* 8, no. 1: 30–37.

Haykal, T., P. Nasr, M. H. Hodroj, R. I. Taleb, R. Sarkis, M. N. Moujabber, and S. Rizk. 2019. *Annona cherimola* seed extract activates extrinsic and intrinsic apoptotic pathways in leukemic cells. *Toxins* 11, no. 9: 506. http://doig.org/10.3390/toxins11090506

Hendawy, O. M., M. A. ELBana, H. A. Abdelmawlla, N. Maliyakkal, and G. Mostafa-Hedeab. 2019. Effect of *Annona squamosa* ethanolic and aqueous leave extracts on aluminum chloride-induced neuroinflammation in albino rats. *Biomedical and Pharmacology Journal* 12, no. 4: 1723–30. https://dx.doi.org/10.13005/bpj/1801

Ibrahim, F., J. Ali, I. Ghassan, and C. Edmont. 2020. Antioxidant activity and total phenol content of different plant parts of Lebanese *Annona squamosa* Linn. *International Journal of Pharmacy and Pharmaceutical Sciences* 12, no. 8: 100–5. http://dx.doi.org/10.22159/ijpps.2020v12i8.36992

Ibrahim, R. Y. M., A. I. Hassan, and E. K. Al-Adham. 2015. The anti-ulcerative colitis effects of *Annona squamosa* Linn. leaf aqueous extract in experimental animal model. *International Journal of Clinical and Experimental Medicine* 8, no. 11: 21861–70. https://www.ncbi.nlm.nih.gov/pubmed/26885156

Ismail, S., N. Hayati, and N. Rahmawati. 2018. Mechanism of action vasodilation *Annona muricata* L. leaves extract mediated vascular smooth muscles. *IOP Conference Series: Earth and Environmental Science* 144, no. 1: 012006. https://doi.org/10.1088/1755-1315/144/1/012006

Jordán-Suárez, O., P. Glorio-Paulet, and L. Vidal. 2018. Microstructure of *Annona muricata* L. leaves extract microcapsules linked to physical and chemical characteristics. *Journal of Encapsulation and Adsorption Sciences* 8, no. 3: 178–93. https://doi.org/10.4236/jeas.2018.83009

Jordán-Suárez, O., P. Glorio-Paulet, and L. Vidal. 2021. Optimization of processing parameters for the microencapsulation of soursop (*Annona muricata* L.) leaves extract: Morphology, physicochemical and antioxidant properties. *Scientia Agropecuaria* 12, no. 2: 161–168. https://dx.doi.org/10.17268/sci.agropecu.2021.018

Justino, A. B., N. C. Miranda, R. F. Rodrigues, M. M. Martins, N. M. Silva, and F. S. Espindola. 2018. *Annona muricata* Linn. leaf as a source of antioxidant compounds with in vitro anti-diabetic and inhibitory potential against α-amylase, α-glucosidase, lipase, non-enzymatic glycation and lipid peroxidation. *Biomedicine and Pharmacotherapy* 100, no. 1: 83–92. https://doi.org/10.1016/j.biopha.2018.01.172

Kumar, A. S., V. Venkatarathanamma, V. N. Saibabu, N. R. Tentu, P. K. Kota, K. S. Kumar, and A. Vijayalakshmi. 2017. Anti-arthritic activity of *Annona squamosa* leaves methanolic extract on adjuvant induced arthritis in rats. *Research Journal of Pharmacology and Pharmacodynamics* 9, no. 2: 46–56. http://doi.org/10.5958/2321-5836.2017.00009.X

Kumar, Y., A. K. Chandra, K. Shruti, and H. P. Gajera. 2019. Evaluation of antidiabetic and antioxidant potential of custard apple (*Annona squamosa*) leaf extracts: A compositional study. *International Journal of Chemical Studies* 7, no. 2: 889–95.

Lee, C. H., T. H. Lee, P. Y. Ong, S. L. Wong, N. Hamdan, A. A. M. Elgharbawy, and N. A. Azmi. 2021. Integrated ultrasound-mechanical stirrer technique for extraction of total alkaloid content from *Annona muricata*. *Process Biochemistry* 109, no. 1: 104–16. https://doi.org/10.1016/j.procbio.2021.07.006

León-Fernández, A. E., E. N. Obledo-Vázquez, M. A. Vivar-Vera, S. G. Sáyago-Ayerdi, and E. Montalvo-González. 2017. Evaluation of emerging methods on the polyphenol content, antioxidant capacity and qualitative presence of acetogenins in soursop pulp (*Annona muricata* L.). *Revista Brasileira de Fruticultura* 39, no. 1: e358. https://doi.org/10.1590/0100-29452017358

López-Córdoba, A. F., E. M. Morales-Valencia, M. M. Pacheco-Valderrama, and A. S. R. Navarro. 2016. Encapsulation of soursop (*Annona muricata*) fruit pulp in calcium alginate hydrogels. *Agronomía Colombiana* 34, no. 1: S1315–8. https://ri.conicet.gov.ar/handle/11336/57339

Mancini, S., L. Nardo, M. Gregori, I. Ribeiro, F. Mantegazza, C. Delerue-Matos, M. Masserini, and C. Grosso. 2018. Functionalized liposomes and phytosomes loading *Annona muricata* L. aqueous extract: Potential nanoshuttles for brain-delivery of phenolic compounds. *Phytomedicine* 42, no. 1: 233–44. https://doi.org/10.1016/j.phymed.2018.03.053

Mannino, G., C. Gentile, A. Porcu, C. Agliassa, F. Caradonna, and C. M. Bertea. 2020. Chemical profile and biological activity of cherimoya (*Annona cherimola* Mill.) and atemoya (*Annona atemoya*) leaves. *Molecules* 25, no. 11: 2612. https://doi.org/10.3390/molecules25112612

Martínez-Solís, J., F. Calzada, E. Barbosa, and M. Valdés. 2021. Antihyperglycemic and antilipidemic properties of a tea infusion of the leaves from *Annona cherimola* Miller on Streptozocin-induced type 2 diabetic mice. *Molecules* 26, no. 9: 2408. https://doi.org/10.3390/molecules26092408

Mulia, K., F. Fauzia, and E. Krisanti. 2019. Freeze-dried chitosan matrices for slow-release of acetogenins extracted from soursop (*Annona muricata* L.) leaves. In *AIP Conference Proceedings* 2175, no. 1: 020031. https://doi.org/10.1063/1.5134595

Neta, M. T. S. L., M. S. Jesus, J. L. A. Silva, H. C. S. Araujo, R. D. D. Sandes, S. Shanmugam, and N. Narain. 2019. Effect of spray drying on bioactive and volatile compounds in soursop (*Annona muricata*) fruit pulp. *Food Research International* 124, no. 9: 70–77. https://doi.org/10.1016/j.foodres.2018.09.039

Nguyen, M. T., V. T. Nguyen, V. M. Le, L. H. Trieu, T. D. Lam, L. M. Bui, L. T. H. Nhan, and V. T. Danh. 2020. Assessment of preliminary phytochemical screening, polyphenol content, flavonoid content, and antioxidant activity of custard apple leaves (*Annona squamosa* Linn.). *Conference Series: Materials Science and Engineering* 736, no. 6: 062012. https://doi.org/10.1088/1757-899X/736/6/062012

Nugraha, A. S., R. Haritakun, J. M. Lambert, C. T. Dillon, and P. A. Keller. 2021. Alkaloids from the root of Indonesian *Annona muricata* L. *Natural Product Research* 35, no. 3: 481–9. https://doi.org/10.1080/14786419.2019.1638380

Ojezele, O. J., M. O. Ojezele, and A. M. Adeosun. 2016. Comparative phytochemistry and antioxidant activities of water and ethanol extract of *Annona muricata* Linn Leaf, seed and fruit. *Advances in Biological Research* 10, no. 4: 230–5. https://doi.org/10.5829/idosi.abr.2016.10.4.10514

Oyekachukwu, A. R., J. P. Elijah, O. V. Eshu, and O. F. C. Nwodo. 2017. Anti-inflammatory effects of the chloroform extract of *Annona muricata* leaves on phospholipase A2 and prostaglandin synthase activities. *Translational Biomedicine* 8, no. 4: 1–8. https://doi.org/10.21767/2172-0479.100137

Passos, T. U., H. A. C. Sampaio, M. O. D. Sabry, M. L. P. Melo, M. A. M. Coelho, and J. W. O. Lima. 2015. Glycemic index and glycemic load of tropical fruits and the potential risk for chronic diseases. *Food Science and Technology* 35, no. 1: 66–73. https://doi.org/10.1590/1678-457X.6449

Phan, H. T., T. T. Nguyen, and P. N. Thi. 2021. Evaluation of the anti-inflammatory effect of *Annona squamosa* L. fruit peel extracts in mouse models of rheumatoid arthritis. *Journal of Microbiology, Biotechnology and Food Sciences* 11, no. 2: e2075. https://doi.org/10.15414/jmbfs.2075

Pinto, N. C. C, J. B. Silva, L. M. Menegati, M. C. M. R. Guedes, L. B. Marques, T. P. Silva, and R. C. N. de Melo. 2017. Cytotoxicity and bacterial membrane destabilization induced by *Annona squamosa* L. extracts. *Anais da Academia Brasileira de Ciências* 89, no. 3: 2053–73. https://doi.org/10.1590/0001-3765201720150702

Rodríguez-Núñez, J. R., E. Campos-Rojas, J. Andrés-Agustín, I. Alia-Tejacal, S. A. Ortega-Acosta, V. Peña-Caballero, T. J. Madera-Santana, and C. A. Núñez-Colín. 2021. Distribution, eco-climatic characterization, and potential growing regions of *Annona cherimola* Mill. (Annonaceae) in Mexico. *Ethnobiology and Conservation*, no. 10: 1–17. https://doi.org/10.15451/ec2020-10-10.05-1-17

Roduan, M. R. M., R. A. Hamid, C. Y. Kqueen, and N. Mohtarrudin. 2019. Cytotoxicity, antitumor-promoting, and antioxidant activities of *Annona muricata in vitro*. *Journal of Herbal Medicine* 15, no. 3: 100219. https://doi.org/10.1016/j.hermed.2018.04.004

Roy, N., and S. Sasikala. 2016. Influence of pre-treatment on secondary metabolites and hypo-glycemic activity of Custard apple (*Annona squamosa*) Peel. *Malaysian Journal of Nutrition* 22, no. 3: 433–42.

Ruiz-Ruiz, J. C., A. J. Matus-Basto., A. J. Acereto-Escoffié, and M. R. Segura-Campos. 2017. Antioxidant and anti-inflammatory activities of phenolic compounds isolated from *Melipona beecheii* honey. *Food and Agricultural Immunology* 28, no. 6: 1424–37. https://doi.org/10.1080/09540105.2017.1347148

Sabapati M., N. Palei, C. Kumar, A. Kumar, and R. B. Molakpogu. 2019. Solid lipid nanoparticles of *Annona muricata* fruit extract: Formulation, optimization and *in vitro* cytotoxicity studies. *Drug Development and Industrial Pharmacy* 45, no. 4: 1–33. https://doi.org/10.1080/03639045.2019.1569027

Sahu, M., N. K. Sahoo, V. Alagarsamy, and B. P. Rath. 2016. Comparative evaluation of anti-diabetic and antioxidant activities of aqueous fruit peel and leaf extracts of *Annona squamosa* on high-fat diet and multiple low dose streptozotocin mouse model of diabetes. *Austin Journal of Pharmacology and Therapeutics* 4, no. 1: 1081–7.

Shehata, M. G., M. M. Abu-Serie, N. M. A. El-Aziz, and S. A. El-Sohaimy. 2021. Nutritional, phytochemical, and in vitro anti-cancer potential of sugar apple (*Annona squamosa*) fruits. *Scientific Reports* 11, no. 1: 1–13. https://doi.org/10.1038/s41598-021-85772-8

Syed-Najmuddin, S. U. F., N. B. Alitheen, M. Hamid, and N. M. A. Nik Abd Rahman. 2017 Comparative study of antioxidant level and activity from leaf extracts of *Annona muricata* Linn obtained from different locations. *Pertanika Journal of Tropical Agricultural Science* 40, no. 1: 119–30.

Vasarri, M., E. Barletta, S. Vinci, M. Ramazzotti, A. Francesconi, F. Manetti, and D. Degl'Innocenti. 2020. *Annona cherimola* Miller fruit as a promising candidate against diabetic complications: An *in vitro* study and preliminary clinical results. *Foods* 9, no. 10: 1350. https://doi.org/10.3390/foods9101350

Veerakumar, S., S. S. D. Amanulla, and K. Ramanathan. 2016. Anti-cancer efficacy of ethanolic extracts from various parts of *Annona squamosa* on MCF-7 cell line. *Journal of Pharmacognosy and Phytotherapy* 8, no. 7: 147–54. https://doi.org/10.5897/JPP2016.0398

Zorofchian-Moghadamtousi, S., E. Rouhollahi, H. Karimian, M. Fadaeinasab, M. Firoozinia, M. A. Abdulla, and H. A. Kadir. 2015. The chemopotential effect of *Annona muricata* leaves against Azoxymethane-induced colonic aberrant crypt foci in rats and the apoptotic effect of acetogenin annomuricin E in HT-29 cells: A bioassay-guided approach. *PLOS ONE* 10, no. 4: e0122288. https://doi.org/10.1371/journal.pone.0122288

CHAPTER 7

Importance of Plant Secondary Metabolites with Biological Activity on Selected Commercial Crops

Rosa Isela Ventura-Aguilar,
A. Berenice Aguilar-Guadarrama,
and Lorena Reyes-Vaquero

CONTENTS

DOI: 10.1201/9781003166535-7

7.1 INTRODUCTION

Crop production is the primary source for obtaining human food; however, large numbers of fungi, bacteria, nematodes, viruses, and insect pests infect important crops, reducing yields and crop survival. Phytopathogenic microorganisms can cause diseases in various plant tissues or organs, such as the leaves, stems, roots, vascular system, and fruits. Plant pathogens use different mechanisms to attack the host, giving rise to specific symptoms (Tawfeeq and Furtado 2020).

During production and storage, plant diseases caused by phytopathogenic microorganisms represent a worldwide economic loss estimated at $220 billion per year (International Plant Protection Convention 2021). Consequently, the use of pesticides has increased; however, this practice pollutes the environment and risks human health. For this reason, biological control strategies are an important research area (He et al. 2019).

Biofungicides are products derived from beneficial microorganisms like bacteria and fungi that inhibit fungal pathogens of crop plants, thereby controlling the diseases normally caused. The activity of biofungicides depends on the mechanism used by the biological organism. For example, they can compete for space in the rhizosphere, act as parasites, generate antibiotics, or produce metabolites which can antagonize the growth of the pathogen or induce defense pathways in the host plant (Bhattacharyya et al. 2016).

Viruses are also causal agents of plant infections, affecting the ecosystems in which they are established. Some, though not all, viruses need an insect vector to move from one plant to another, so control of the vector can help control the virus (Jackson and Chen-Charpentier 2018).

Biopesticides (biochemical pesticides) are derived from natural products, including potassium bicarbonate, plant extracts, and essential oils. Notably, plant extracts, containing secondary metabolites, could be an excellent option for developing biopesticides against phytopathogens because they are affordable, have ecological and sustainable potential for their use in plant and crop protection, and use several processes focused on the efficient extraction of specific chemical compounds (Marrone 2019).

In addition, plant extracts have specific toxicity to control microorganisms and pests without destroying beneficial organisms. This is because secondary metabolites act like the defenses used by the plant to protect against attack by external agents; hence, their biosynthesis is increased only under conditions of stress. This chapter will provide an analysis of plant secondary metabolites with biological activity toward phytopathogenic microorganisms that affect crops of commercial importance.

7.2 SECONDARY METABOLITES DERIVED FROM PLANTS

Secondary metabolites (SMs) are a broad group of chemical compounds of low molecular weight, derived from the primary metabolism of plants (Chiocchio et al. 2021). The biosynthesis and accumulation of SMs depend on environmental conditions and the phenotype and genotype of plants. Thus, changes in SMs are observed even within the same species growing under different environmental conditions. Furthermore, SMs accumulate in different plant tissues and have several functions, such as defense against herbivores, fungi, and bacteria (Khare et al. 2020).

According to their biosynthetic pathways and chemical structures, SMs are classified into three main groups: phenolics compounds, terpenes, and alkaloids. Phenolic compounds are precursors of aromatic molecules produced in plants by the shikimic acid and malonic acid pathways. These are classified as simple phenols, phenolic acids, flavonoids, or tannins, based on the number of aromatic rings, carbon atoms, and hydroxyl groups (Chiocchio et al. 2021). The shikimate pathway is the primary source of phenylpropanoids and other aromatic compounds, and the intermediate *p*-coumaroyl-CoA is the precursor of coumarins, flavonoids, lignans, stilbenes, and catechins (Chandran et al. 2020).

On the other hand, the mevalonic acid (MVA) and the 2-C-methylerythritol 4-phosphate (MEP) pathways in the cytosol and plastids, respectively, of plant cells are precursors of terpenes. Through the MVA pathway, compounds such as brassinosteroids, sesquiterpenoids, phytosterols, triterpenoids, and polyprenols are synthesized. In contrast, the MEP pathway gives rise to monoterpenoids, diterpenoids, sesquiterpenoids, tocopherols, and plastoquinones (Chandran et al. 2020); terpenoids are classified according to the number of isoprene units in their molecule (Khare et al. 2020). Other precursors of terpenoids are pyruvate and glyceraldehyde-3-phosphate responsible for isopentenyl pyrophosphate (IPP) and dimethylallyl pyrophosphate (DMAPP) biosynthesis. Finally, alkaloids are molecules with atoms of nitrogen in their structure and are derived from amino acids such as lysine, tyrosine, and tryptophan. They are grouped as monoterpenoid indole, benzylisoquinoline, tropane, purine, acridone, furoquinolines, and pyrrolizidine alkaloids (Khare et al. 2020).

In addition, SMs have diverse mechanisms of action against phytopathogenic agents, including fungi, bacteria, nematodes, and host plant cells infected by a virus. The response includes inhibition or denaturation of proteins, inhibition of protein or DNA synthesis of egg hatching, spore germination, disruption of cellular components, mortality, and hyphal modifications. These responses and other effects depend on the type of botanical compound and phytopathogen involved (Lengai et al. 2020).

Plants synthesize diverse SMs, protecting them against predators and pathogenic microorganisms (Fig. 7.1). Zaynab et al. (2018) reported that metabolites such as limonene, thymol, citronellal, gossypol, and demissine showed larvicidal activity against pests such as *Spodoptera litura*, *Atta cephalotes*, *Heliothis virescens*, *Heliothis zea*, and *Leptinotarsa decemlineata*. On the other hand, it has been reported that metabolites such as carvacrol, thymol, or citral inhibit the growth of fungi such as *Botrytis cinerea*, *Rhizoctonia solani*, *Fusarium moniliforme*, *Sclerotinia sclerotiorum*, and *Penicillium italicum* (Poveda et al. 2020). SMs with antifungal activity include phenolic compounds, terpenes, saponins, and alkaloids (Lengai et al. 2020).

This wide diversity of chemical structures that characterize the SMs derived from plants makes them interesting compounds for various industrial sectors. However, the extraction of these compounds is limited by the availability of plant material, the nature and concentration of these compounds, and the effect of abiotic factors on their extraction process, with *in vitro* culture offering an alternative form of production (Espinosa-Leal et al. 2018).

7.3 EXTRACTION OF SECONDARY METABOLITES FROM PLANTS

SMs are extracted from plants by different chemical procedures based on the chemistry of the SM and the polarity of the solvent, and sometimes heat is needed (Lefebvre et al. 2021). Basically, the preparation of extracts to obtain the SM starts by drying and milling the plant material. Thereafter, non-polar, medium-polar, and high-polar solvents are used to assure complete extraction, this being one of the most used techniques because it preserves the anti microbial properties in the extracts (Lefebvre et al. 2021). Tembo et al. (2018) reported that extracts of *Bidens pilosa*, *Lantana camara*, *Lippia javanica*, *Tephrosia vogelii*, *Tithonia diversifolia*, and *Vernonia amygdalina* had pesticidal activity against the adults and larvae of spiders, ladybeetles, and adults of hoverflies. To prepare the extracts, the authors blended 1 kg of powdered dry tissue of each plant species with 10 L of water and kept them at room temperature (20±5 °C) for 24 h. To increase

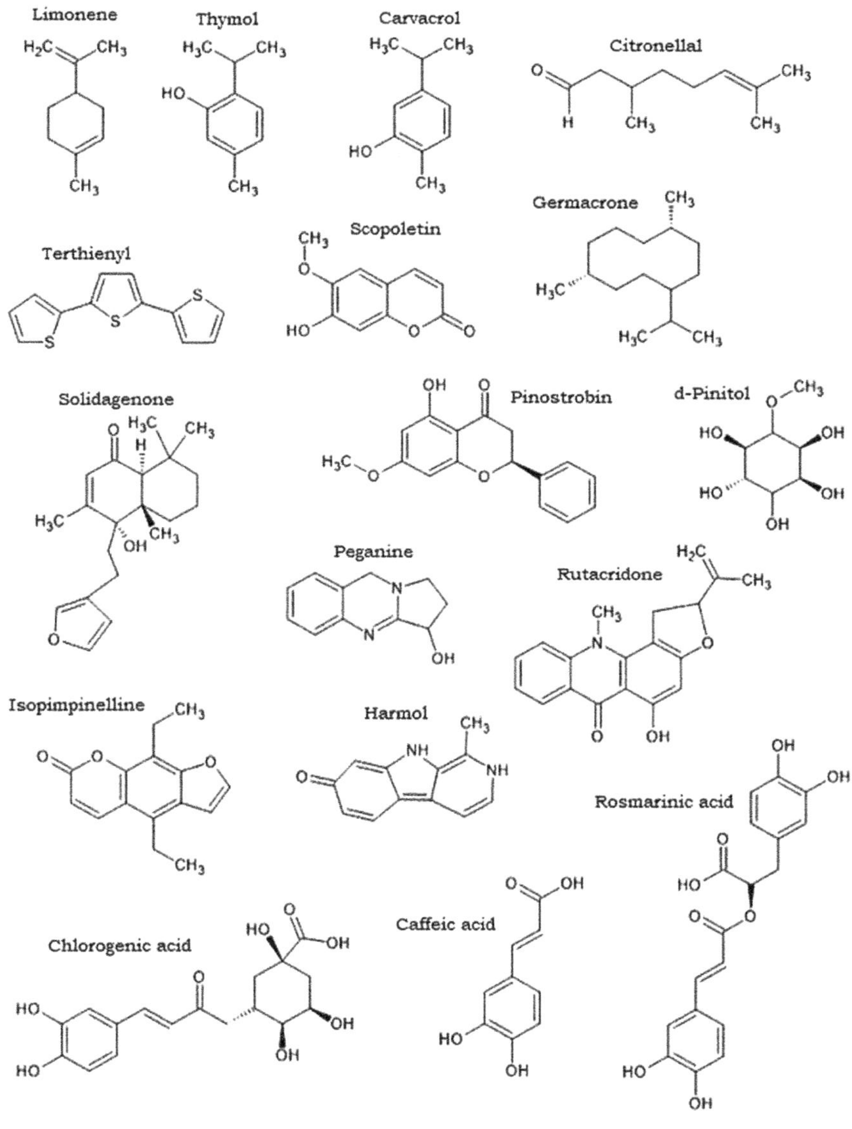

FIGURE 7.1 Example of plant secondary metabolites with biological activity against plant-pathogenic microorganisms.

the efficiency of extracting non-polar compounds, 0.1% of detergent was included. Finally, the suspension was filtered, and the plant material was discarded. The extract was used directly on the test plants.

Some reports find that camphor is the major component with insecticidal activity in *L. javanica*, while caryophyllene, camphene, cymene, eucalyptol, linalool, thymol, α-pinene, *Z*- and *E*-α-terpineol, 2-carene, and α-cubebene are the minor components (Tembo et al. 2018). From *T. vogelii*, deguelin, tephrosin and rotenone were identified previously, whereas germacrene D, limonene, α-phellandrene, β-caryophyllene, and 1,8-cineole were identified in *L. camara*. β-caryophyllene has been reported to have antifungal activity. Finally, the main insect repellent components of *V. amygdalina* identified were vernodalin, 11,13-dihydrovernodalin, and several vernoniosides. The authors found that pesticides extracted from plants can successfully control pests in legume crops, achieving yields similar to those obtained using synthetic pesticides (Tembo et al. 2018).

In another case study, 17 plants were assayed against phytopathogenic microorganisms including *Alternaria alternata*, *Aspergillus flavus*, *Aspergillus niger*, *B. cinerea*, *Bipolaris oryzae*, *Curvularia lunata*, *Curvularia oryzae*, *Fusarium solani*, *F. oxysporum*, *Macrophomina* sp., and *Trichoderma* sp. Sequential extraction using hexane, chloroform, and ethanol was performed on each plant species. The extracts obtained of 17 plants, the hexane extract of *Blumea mollis* was the only one which had activity against the tested fungi. Twenty-seven compounds in the hexane extract from *B. mollis* were detected by gas chromatography–mass spectrometry (GC–MS) the relative abundance of main compounds were: 4-methylene-1-methyl-2-(2-methyl-1-propen-1-yl)-1-vinyl (34.21%), caryophyllene oxide (26.73%) and bicyclo[7.2.0]undec-4-ene, 4,11,11-trimethyl-8-methylene-[1*R*-(1*R**,4*Z*,9*S**)] (18.24%); it was suggested that such compounds in the hexane extract from *B. mollis* are responsible for antifungal activity (Sivanandhan et al. 2018).

Extracts from the leaves of *Artocarpus heterophyllus* were tested as an antifungal agent against *Colletotrichum gloeosporioides* and *Penicillium italicum* and extracted by each of three different methods: microwave (MAE), high hydrostatic pressure (HHP), and ultrasonic-assisted extraction (UAE). The authors found that 2 mg/mL of methanolic extract obtained by HHP inhibited 83.78% of mycelial growth of *C. gloeosporioides*; phenolic compounds such as 3, 5-dicaffeoylquinic acid and 1-isoleucine-pentafluoropropionic-pentadecyl ester have been reported that could be responsible for this antifungal activity (Vázquez-González et al. 2020).

Aguilar-Veloz et al. (2021) reported that a hydroalcoholic (4:1) extract obtained by MAE recovered phenolic compounds such as apigenin C-glycosides, apigenin-6,8-C-diglucoside, apigenin 8-C-xyloside-6-C-glucoside, chlorogenic, isoschaftoside, isorhamnetin glycoside, quercetin glucoside-*O*-rutinoside, 3,5-dicaffeoylquinic and *cis*-5-*O*-*p*-coumaroylquinic acids from *A. heterophyllus* leaves. This extract reduced *Alternaria* sp. growth.

7.4 DISEASES CAUSED BY PHYTOPATHOGENIC MICROORGANISMS IN HIGH-VALUE-ADDED CROPS

A plant pathogen is any microbial agent that prevents a plant from performing to its maximum potential and, in general, is caused by nematodes, fungi, bacteria, or viruses; insects, on the other hand, act as crop pests. Plant pathogens can infect various types of organs and tissues such as leaves, roots, stems, shoots, seeds, fruits, or vascular tissues (Kaur et al. 2019).

A plant-parasitic nematode genus, such as *Pratylenchus*, *Radopholus*, *Hirschmanniella*, *Anguina*, *Bursaphelenchus*, *Ditylenchus*, *Heterodera*, *Globodera*, *Meloidogyne*, and *Aphelenchus*, can cause damage to plants that encourages other organisms such as fungi, bacteria and viruses to penetrate the plant and cause various diseases (Smant et al. 2018). Nematode damage results in about $100 billion crop losses worldwide (Ghareeb et al. 2020). Nematodes can affect any agricultural crop, including bananas, coconuts, cereals, legumes, potatoes, sugarcane, sugar beet and sweet potatoes, among others (Table 7.1). For this reason, the production and quality of crops are affected, causing economic losses. In addition, one more problem is added, since nematodes act as vectors for the transmission of viral diseases (Mesa-Valle et al. 2020). An example of nematodes is the root-knot nematode (*Meloidogyne* spp.) which is one of the most damaging plant-parasitic nematode genera for crops, causing great economic losses (Rashid et al. 2021). Cyst nematodes (*Globodera* spp. and *Heterodera* spp.) cause losses of up to 30% in the yield of wheat, barley, and oat crops. Similarly, *Nacobbus aberrans* reduces crop yield in tomatoes, potatoes, beans, and sugar beets in countries such as Argentina, Bolivia, Chile, Ecuador, Mexico, Peru, and the USA. (Velasco-Azorsa et al. 2021). Finally, *Aphelenchus* is a genus of plant-parasitic foliar nematodes that mainly affects chrysanthemum (Mandal et al. 2021).

Another critical group of microorganisms in plants are fungi. Phytopathogenic fungi are responsible for nearly 30% of all crop diseases, affecting both pre- and postharvest stages (Raveau et al. 2020). Depending on the mechanism of infection, fungi can be grouped into biotrophs, necrotrophs, and hemibiotrophs (Shuping and Eloff 2017). Some of the most economically devastating fungi are *Magnaporthe oryzae*, affecting rice and wheat, *B. cinerea*, *Colletotrichum* spp. and *Fusarium* spp., which

TABLE 7.1 Secondary Metabolites as Biological Control Agents in Crops of Commercial Interest

Plant Material	Secondary Metabolites	Microorganism to Control	Crops Affected	Disease Symptoms	ReferenceS
Brassica juncea and *Lupinus* sp.	Not reported.	*Pratylenchus penetrans* and *Meloidogyne chitwoodii.*	It has a broad host range of nearly 400 plant species, such as maize, potato, cereals, and several vegetables.	*P. penetrans* enters and leaves the host roots during their life cycle, moving actively to feed and reproduce. The symptoms include plant dwarfing, early senescence, and wilting. On potatoes, root-lesion nematodes penetrate through the lenticels and migrate to the adjacent tissues forming lesions with an appearance of cross-shape marks or protuberances.	Cóndor 2019; Figueiredo et al. 2021
Cymbopogon nardus and *Dysphania ambrosioides.*	Ascaridole and citronellal are the primary metabolites that cause 40% to 97% mortality.	*Meloidogyne incognita*	Citrus, banana, tomato, watermelon, pepper, carrots, lettuce, onions, soybean, grapes, potato, beetroot, peanuts, spinach, dry bean, eggplant, and maize.	Large galls or "knots" can form throughout the root system of infected plants.	De Freitas et al. 2020; Sithole et al. 2021

(*Continued*)

TABLE 7.1 (CONTINUED) Secondary Metabolites as Biological Control Agents in Crops of Commercial Interest

Plant Material	Secondary Metabolites	Microorganism to Control	Crops Affected	Disease Symptoms	ReferenceS
Cucurbita maxima, *Prunus africana*, *Pelargonium sidoides*, *Croton sylvaticus*, *Solanum aculeastrum*, *Vernonia colorata*, *Searsia lancea*, and *Merwilla plumbea*.	Alkaloids, terpenoids, phenolics, and flavonoids of *Cucurbita maxima* inhibited *Meloidogyne* spp. by 77% to 96% after 72 h of exposure.	*Meloidogyne* spp. (root-knot nematode).	Cucurbits, tomatoes, eggplants, and lettuce.	Formation of typical galls on the roots, ranging in size from 1 to 10 mm in diameter. Nutrient and water uptake are substantially reduced because of the damaged root system, resulting in weak and low-yielding plants, or death in cases of high infestation.	Djian-Caporalino 2012; Sithole et al. 2021
Thymus vulgaris and *Zingiber officinale*.	Thymol, carvacrol, β-cymene, and α-terpinolene are present in *T. vulgaris* extract (recommended dose of 8 mg/mL). Gingerol, cedrene zingiberene and α-curcumene present in *Z. officinale* extract inhibited the phytopathogenic fungi at 16 mg/mL.	*Fusarium oxysporum*, *Pythium aphanidermatum*, and *Rhizoctonia solani*.	Tomato, potato, apple, rice, barley, soybean, corn, and sugar beet.	Vascular wilt, root rot, stem cankers, and tuber blemishes.	Al-Rahmah et al. 2013

(*Continued*)

TABLE 7.1 (CONTINUED) Secondary Metabolites as Biological Control Agents in Crops of Commercial Interest

Plant Material	Secondary Metabolites	Microorganism to Control	Crops Affected	Disease Symptoms	ReferenceS
Phenolics obtained commercially.	The inhibition ratio of carvacrol and thymol was higher than 90% against *Botrytis cinerea* at 50 µg/mL. The compounds without methyl or isopropyl groups (*o*-cresol, *m*-isopropylphenol and 2,5-xylenol) showed lower antifungal activity than carvacrol and thymol.	*Phytophthora capsici*, *Phytophthora nicotianae*, *Alternaria solani*, *Botrytis cinerea*, *Fusarium oxysporum*, *Pyricularia grisea* and *Rhizoctonia solani*.	Potato, tomato, apple, cereals, fruit trees, strawberry, berries, cucumber, pepper, eggplant, rose, and gerbera.		Wang et al. 2019
Eucalyptus tereticornis, *Xanthium sibiricum*, *Artemisia argyi*, *Tupistra chinensis*, and *Pyrola calliantha*.	Phenolics and flavonoids in 95% ethanol and ethyl acetate extracts of *T. chinensis* rhizomes might be good antifungal agents controlling *Rhizopus stolonifer* and *B. cinerea* in postharvest fruits.	*Aspergillus niger*, *Botrytis cinerea*, *Penicillium digitatum*, *Penicillium expansum*, *Penicillium italicum*, and *Rhizopus stolonifer*.	Fruits and vegetables.	Postharvest rots and bunch rot.	Yao et al. 2021

(*Continued*)

TABLE 7.1 (CONTINUED) Secondary Metabolites as Biological Control Agents in Crops of Commercial Interest

Plant Material	Secondary Metabolites	Microorganism to Control	Crops Affected	Disease Symptoms	ReferenceS
Carum copticum, *Mallotus philippensis*, *Citrullus colocynthis*, *Calotropis procera*, *Embelia ribes*, *Ricinus communis*, *Lawsonia inermis*, *Amomum subulatum*, *Operculina turpethum*, and *Santalum album*.	*Lawsonia inermis*, *Embellia ribes*, and *Santalum album* showed antibacterial activity against all the evaluated bacteria.	*Escherichia coli*, *Bacillus cereus*, and *Staphylococcus aureus*.	Carrot, zucchini, pepper, lettuce, papaya, cucumber, garlic, tomato, and onion.	These are not phytopathogenic bacteria.	Hussain et al. 2011

(*Continued*)

TABLE 7.1 (CONTINUED) Secondary Metabolites as Biological Control Agents in Crops of Commercial Interest

Plant Material	Secondary Metabolites	Microorganism to Control	Crops Affected	Disease Symptoms	ReferenceS
Larrea tridentata, *Flourensia cernua*, *Lippia graveolens*, *Agave lechuguilla*, *Yucca filifera*, *Opuntia ficus-indica*, and *Carya illinoensis*.	Lanolin extracts promoted the highest foodborne bacterial growth inhibition. Creosote bush extracts had the highest polyphenols content and promoted bacterial growth inhibition.	*Enterobacter aerogenes*, *Escherichia coli*, *Salmonella typhi*, and *Staphylococcus aureus*.	Lettuce, tomato, chili, onion, lettuce, arugula, spinach, and cilantro.	These are not phytopathogenic bacteria.	Luna-Guevara et al. 2019; Mendez et al. 2012
Coffea arabica	From 5 to 20 mM of caffeine is necessary to achieve the minimum inhibitory concentration, and from 43 to 100 mM for the minimum bactericidal concentration. Caffeine increased the bacterial generation time of all tested species and caused changes in the cell morphology of bacteria.	*Ralstonia solanacearum*, *Clavibacter michiganesis* ssp. *sepedonicus*, *Dickeya solani*, *Pectobacterium atrosepticum*, *Pectobacterium carotovorum* ssp. *carotovorum*, *Pseudomonas syringae*, and *Xanthomonas campestris*.	Tomato, potato, cucumber, cherry, plum, apricot, almond and pear trees, among others.	*Pectobacterium carotovorum* cause soft rot disease of various plant hosts and blackleg in potato by degradation of the plant cell wall. For *Pseudomonas* spp., the symptoms expressed by their hosts are blighting of dormant buds, resulting in the formation of small cankers on the branch and stem cankers, oozing or gummosis, branch dieback, and occasionally complete tree death.	Sledz et al. 2015; Shin-Ichi 2014; Khumbuzile et al. 2020

all have a broad host range, and *Puccinia* spp. affecting wheat (Almeida et al. 2019).

Kaur et al. (2019) reported that some fungal diseases cause mold, rust, mildew, rot, canker, spot, anthracnose, and wilt. These diseases affect crops of commercial interest, such as cotton, groundnut, citrus, and oil palm, vegetables, such as chili, cucumbers, and tomato, grains, such as corn, rice, soybean, bean, pea, and wheat, among others (Table 7.1).

Fungal diseases manifest themselves through different symptoms and as generalists or specialists on particular hosts; for example, *B. cinerea* is the causal agent of gray mold, affecting 586 plant genera. The main problem caused by *B. cinerea* is the infection of flowers and fruits after harvest, producing fruit rot in strawberry and stem canker in cucumber, pepper, and eggplant, whereas, in rose and gerbera, the petals are infected (Bardin and Gullino 2020). The genus *Alternaria* is a fungus that is a causal agent of leaf spots and infects vegetables, cereals, and fruit during production and storage (Freire et al. 2017). Another disease caused by this genus is early blight, affecting crops like tomatoes and potatoes (Jain et al. 2019).

Sclerotinia sclerotiorum is a necrotrophic fungal pathogen that causes Sclerotinia stem rot (SSR) in the oilseed crop *Brassica napus*. On the other hand, *Rhizoctonia solani* is considered one of the most destructive fungal pathogens of rice, leading to a loss in crop yield that ranges between 20% and 40% (Pradhan et al. 2020). *F. oxysporum* elicits a severe vascular wilt disease in crops such as potato, tomato, and other Solanaceae members (Jain et al. 2019). Fungi causing rusts, like *Uromyces dianthi*, *Phragmidium mucronatum*, *Phragmidium tuberculatum*, and *Puccinia horiana*, affect ornamental crops, such as carnation, rose and chrysanthemum (Bardin and Gullino 2020). Anthracnose is the most common name given to diseases caused by *Colletotrichum* spp., which affects fruit crops, strawberries, peaches, apples, grapes, blueberries, and cranberries (Dowling et al. 2020).

After harvesting, fruits and vegetables are susceptible to spoilage and contamination as a result of inappropriate handling and the growth of microbes, such as fungi and bacteria (Kumar and Iqbal 2020). Microbial growth can occur during production, harvest, handling, storage, packaging, marketing, and after purchase by the consumer. Fungi such as *B. cinerea*, *Penicillium* spp., *Fusarium*, *Geotrichum*, *Colletotrichum*, and *Aspergillus* are causal agents for the decay of fruit and vegetables in the postharvest stage (Shuping and Eloff 2017).

On the other hand, bacteria penetrate the plant surface through natural openings and wounds, causing diseases like soft rots, spots and wilts in different crops (Kaur et al. 2019; Catara and Bella 2020). *Dickeya* and *Pectobacterium* cause bacterial soft rot, which is characterized by the dissolution of host tissues. It causes severe damage in vegetable or ornamental plant production, such as potatoes, sugar beets, broccoli, and cauliflower. Also, these bacteria cause blackleg of potatoes, foot rot in rice, and bleeding canker of pear (Charkowski 2018). Bacterial wilt affects the production of up to 250 plant species worldwide, such as tobacco, tomato, potato, banana, eggplant, pepper, sunflower, and other solanaceous plants (Jiang et al. 2017; Catara and Bella 2020). The bacterial wilt in the Cucurbitaceae is caused by *Erwinia tracheiphila* and affects *Cucumis* spp. (cucumber and muskmelon) and *Cucurbita* spp. (pumpkin, squash, and yellow-flowered gourds) as reported by Catara and Bella (2020).

Tomato bacterial canker is one of the most severe tomato diseases, and it is caused by *Clavibacter michiganensis* (Catara and Bella 2020). Another disease that affects members of the Cucurbitaceae is bacterial fruit blotch. This disease affects leaves and fruits and is caused by *Acidovorax citrulli* (Zivanovic and Walcott 2017).

Bacterial spot is a disease caused by *Xanthomonas* lineages, such as *X. euvesicatoria* pv. *Euvesicatoria*, *X. euvesicatoria* pv. *Perforans*, *X. hortorum* pv. *Gardneri*, and *X. vesicatoria*, affecting tomato and pepper crops; for example, the heaviest crop losses on pepper result from the shedding of blossoms and young developing fruits (Catara and Bella 2020).

According to Moriones and Verdin (2020), infections and diseases caused by viruses in plants lead to significant yield loss and threaten crop production. Some viral diseases are mottling, distortion, and dwarfing (Kaur et al. 2019). The plant-pathogenic viruses need vectors to reach their hosts, mainly insects. Solanaceae crops, such as pepper, tomato, and potato, are affected by potyvirus, cucumovirus, alfamovirus, fabavirus, polerovirus, orthophotovirus, begomovirus, crinivirus, tobamovirus, and ilarvirus. These viruses are transmitted by aphids and whiteflies (Moriones and Verdin 2020).

On the other hand, insect pests cause crop losses worldwide through direct damage and plants diseases. The chewing insects feed on the plant's surface, causing damage to different parts of the plants. The larvae of some chewing insect species live within the plant tissues and may bore their way into the plant; these groups include the leaf miners, stem borers, and root borers (Douglas 2018). Some examples of pests in commercial crops are the carrot psyllid (*Trioza apicalis*) and the carrot rust fly (*Psila rosae*) (Cotes et al. 2018).

The American serpentine leafminer, *Liriomyza trifolii*, affects vegetable crops such as tomato and pepper in Europe. The peach-potato aphid, *Myzus persicae*, the cabbage aphid, *Brevicoryne brassicae*, and the silver leaf whitefly, *Bemisia tabaci*, are important pests of vegetable crops, causing severe direct and indirect damage. The onion thrips, *Thrips tabaci*, and the western flower thrips, *Frankliniella occidentalis*, are significant pests of several vegetable and ornamental crops worldwide, and these species are also vectors of significant viruses, such as the tomato spotted wilt virus (Karkanis and Athanassiou 2020).

In potato, eggplant, and tomato crops, the Colorado potato beetle, *Leptinotarsa decemlineata*, is one of the most destructive pests (Tajmiri et al. 2017). As can be seen, the number of microorganisms that affect crops of interest are many and varied, so strategies are required for their integrated management.

7.5 SECONDARY METABOLITES USED FOR THE PROTECTION OF CROPS OF COMMERCIAL INTEREST

Pests and diseases of crops are responsible for losses of 40% in crop production worldwide (International Plant Protection Convention, 2021). The strategy for conventional disease management is based on the use of synthetic chemical pesticides. However, the excessive use of these products has become less effective and more restricted due to the emergence of resistant pathogens and the adverse environmental effects resulting from the inappropriate use of these chemicals (Abubakar et al. 2020); in addition, the use of synthetic pesticides negatively affects beneficial microorganisms (Tembo et al. 2018). Thus, alternatives are being sought among plant extracts for managing nematodes, fungi, bacteria, and pest insects since plant extracts contain more than one compound with biological activity against phytopathogenic organisms (Shuping and Eloff 2017).

Metabolites with activity against important crop pests are found in plants of the families Myrtaceae, Lauraceae, Rutaceae, Lamiaceae, Asteraceae, Apiaceae, Cupressaceae, Poaceae, Zingiberaceae, Piperaceae, Liliaceae, Apocynaceae, Solanaceae, Caesalpinaceae, and Sapotaceae (Lengai et al. 2020). For instance, sesquiterpenes are found in the family Solanaceae, stilbenes in the Vitaceae, isoflavones in the Fabaceae, sulfur-based glucosinolates and the associated enzyme myrosinase in the Brassicaceae, limonoids among members of the Meliaceae family, and furanocoumarins in the Rutaceae (Cluzet et al. 2020).

Extracts from various plants have been reported to have nematicidal, nematostatic and egg hatching inhibition activity against plant-pathogenic nematodes. Vázquez-Sánchez et al. (2018) reported that extracts of *Verbesina sphaerocephala*, *Cosmos sulphureus*, and *Senecio salignus* showed nematostatic activity against *Nacobbus aberrans*, whereas extracts of *Witheringia stramoniifolia*, *Tagetes lunulata*, *S. salignus*, and *Lantana camara* caused 73% mortality against *N. abberrans*.

There are other plants that have nematicidal and nematostatic activity against *N. aberrans*; for example, extracts of *Galium mexicanum* and *Alloispermum integrifolium* caused a paralysis effect, and extracts of *Adenophyllum aurantium*, *A. integrifolium*, *Acalypha subviscida*, *G. mexicanum*, and *Heliocarpus terebinthinaceus* resulted in 90% mortality. The nematostatic and nematicide activity of *Adenophyllum aurantium* is due to metabolites like stigmasterol, β-sitosterol, and α-terthienyl (Velasco-Azorsa et al. 2021).

Sithole et al. (2021) evaluated the effect of crude extracts of *Cucurbita maxima*, *Prunus africana*, *Pelargonium sidoides*, *Croton sylvaticus*, *Solanum aculeastrum*, *Vernonia colorata*, *Searsia lancea*, and *Merwilla plumbea* on egg hatching inhibition of *M. incognita*. The authors reported that the extract of *C. maxima* inhibited the egg mass hatching by 77–96% and % inhibition was inversely proportional to the concentration and exposure time.

Rashid et al. (2021) evaluated different fractions of root extracts from *Curcuma longa* against *M. incognita* and reported that the chloroform fraction showed mortality of 25%, whereas extracts obtained by sonication have shown mycelial growth inhibition of between 30% and 78% against plant-pathogenic fungi like *Sclerotinia sclerotiorum*, *F. oxysporum*, *F. graminearum*, *Fusarium chlamydosporum*, *Fusarium tricinctum*, *F. culmorum*, *Rhizopus oryzae*, *Cladosporium cladosporioides*, *A. alternata*, *B. cinerea*, and *Colletotrichum higginsianumi* (Chen et al. 2018). The antifungal effects of extracts from *C. longa* were related to the disruption of fungal cell membrane systems and the inhibition of ergosterol synthesis and the respiratory chain; this activity is due to metabolites, such as curdione, curcumin, isocurcumenol, curcumenol, curzerene, β-elemene, germacrone, and curcumol present in the extract (Chen et al. 2018).

Other studies have reported the biological activity of metabolites from plants against plant-pathogenic fungi. Sivanandhan et al. (2018) reported that a hexane extract of leaves from *Blumea mollis* showed maximum zone growth inhibition against *A. flavus*, *A. niger*, *B. cinerea*, *Curvularia oryzae*,

C. lunata, *Fusarium solani*, and *F. oxysporum*. The authors identified the main metabolites as cycloheptane, 4-methylene-1-methyl-2-(2-methyl-1-propen-1-yl)-1-vinyl, caryophyllene oxide, and isocaryophyllene.

Bayar and Yilar (2019) evaluated the antifungal activity of methanolic and hexanic extracts of *Salvia virgata* against *R. solani*, *A. solani*, *Fusarium oxysporum* sp. *radices-lycopersici*, and *Verticillium dahliae*. The authors reported that the methanol extract inhibited the mycelial growth on *V. dahliae* and *F. oxysporum* sp. *radices-lycopersici* at rates of 38.77% and 28.17%, respectively, but had no effect on the mycelial growth of *A. solani* and *R. solani*. In contrast, the hexane extract inhibited the mycelial growth of *R. solani*, *A. solani*, *F. oxysporum* sp. *radicis-lycopersici*, and *V. dahliae* at the rates of 72.22%, 37.01%, 36.04% and 2.43%, respectively.

Daradka et al. (2021) reported the fungicidal activity in ethanolic extracts of leaves from ten plants, namely *Bupleurum falcatum*, *Citrullus colocynthis*, *Dodonaea viscosa*, *Ficus palmata*, *Olea chrysophylla*, *Otostegia fructicosa*, *Psiadia arabica*, *Pulicaria crispa*, *Rumex vesicarius*, and *Zygophyllum simplex*, against *A. alternata* and *F. oxysporum*. The authors identified compounds in the plant extracts such as alkaloids, flavonoids, saponins, tannins, and terpenoids, and reported that extracts from *Pulicaria crispa* and *Olea chrysophylla* were the most effective in inhibiting mycelial growth, ranging from the maximum zone of inhibition of 10 to 29 mm against *A. alternata* and 12 to 31 mm for *F. oxysporum*.

Reyes-Vaquero et al. (2021) reported that leaf and root ethanolic extracts of *Ruta graveolens* obtained by the sonication method inhibited 69% to 73% growth of *F. oxysporum*, *F. proliferatum* and *Stemphylium vesicarium*. Fatty acids, alkaloids, and terpenes were the most abundant metabolites in leaves, and alkaloids in roots. Specifically, the antifungal activity correlated with the presence of compounds such as skimmianine, campesterol, phytol, linolenic acid, oleic acid, palmitic acid, linoleic acid, dodecanoic acid, isopimpinelline, rutacridone, 1-hydroxy-10-methyl-9 (10*H*)-acridinone, and xanthotoxin identified in the extracts of the aerial parts and roots.

Pham et al. (2017) reported that ethyl acetate and dichloromethane extracts of *Rheum tanguticum* rhizomes showed potent efficacy to control plant fungal diseases caused by *Magnaporthe oryzae*, *Phytophthora infestans*, *Puccinia recondita*, *Blumeria graminis* f. sp., and *Colletotrichum coccodes*, and inhibited growth of the bacteria *Burkholderia glumae*, *Clavibacter michiganensis* subsp. *michiganensis*, *Pseudomonas syringae* pv. *lachrymans*, and *Ralstonia solanacearum*. They also reported that metabolites, such as physcion, inhibited the development of barley powdery mildew by 80%, whereas chrysophanol showed greater efficacy, with values of 95%.

On the other hand, there are several studies on plant extracts against plant-pathogenic bacteria. For example, El-Hefny et al. (2017) reported that *n*-butanol extracts of *Callistemon viminalis* flowers and *Eucalyptus camaldulenses* bark showed the maximum inhibition zones against colony growth of *Ralstonia solanacearum* (18 mm and 20 mm, respectively) and *Pectobacterium carotovorum* subsp. *carotovorum* (16 mm for both plant samples).

Mori et al. (2019) reported that *Erwinia chrysanthemi* growth was inhibited by ethanol extracts of *Phyllostachys heterocycla*. This extract also exhibited an inhibitory effect on the growth of fungal phytopathogens such as *B. cinerea*, *Glomerella cingulata*, *Trichoderma harzianum*, and *Hypocrea lactea*.

Shaheena and Issa (2020) evaluated the total alkaloid extract of *Peganum harmala* seeds against *Ralstonia solanacearum* Phylotype II, *Erwinia amylovora*, *Pectobacterium carotovorum* subsp. *carotovorum*, and *Burkholderia gladioli*. The results showed that *P. harmala* seeds contained alkaloids, including egamine, harmol and harmine, which inhibited the *in vitro* growth of *R. solanacearum*, *E. amylovora*, *P. carotovorum* subsp. *carotovorum*, and *B. gladioli*, and has the ability to reduce brown rot disease caused by *R. solanacearum* in tubers of potato plants *in vivo* by 50%.

As can be seen, some plants contain metabolites, such as alkaloids and terpenes, that can be used to control plant diseases caused by fungi and bacteria.

Some research has also been published regarding the use of plant extracts to control crop pests. Reyes-Vaquero et al. (2016) reported that methanolic extracts obtained by maceration of leaves from *Bougainvillea glabra* "Variegata" killed 85% of fall armyworm larvae, causing incomplete ecdysis and malformation in larvae and pupae, with no adults emerging. Furthermore, Evangelista-Lozano et al. (2018) reported that D-pinitol was the main active metabolite identified in the methanolic extracts and was also responsible for larvicidal activity against second instar larvae of *Spodoptera frugiperda*, reducing larvae weight, causing the death of larvae, and preventing pupal development.

Khan et al. (2017) evaluated the insecticidal activity of *Daphne mucronata*, *Tagetes minuta*, *Calotropis procera*, *Boenninghausenia albiflora*, *Eucalyptus sideroxylon*, *Cinnamomum camphora*, and *Isodon rugosus* extracts against pea aphids, fruit flies, red flour beetles, and armyworms, which are significant agricultural pest insects, and reported that all the plant extracts tested caused 100% mortality of pea aphids.

Tembo et al. (2018) evaluated the effect of aqueous extracts of fresh leaves from six plant species, *Tephrosia vogelii*, *Vernonia amigdalina*, *Lippia javanica*, *Tithonia diversifolia*, *Bidens pilosa*, and *Lantana camara*, on reducing pest insect abundance on bean, cowpea, and pigeon pea crops. The extracts of *L. javanica*, *T. vogelii*, and *T. diversifolia* reduced pest insect numbers, and had similar effects to synthetic pesticides; additionally, the extracts did not affect beneficial insects.

Furthermore, ethanolic extracts of leaves from *Ocimum gratissimum* elicited chronic mortality of *S. littoralis* larvae, and the mortality increased over time and with increasing extract concentration, as reported by Benelli et al. (2019). Carvacrol, thymol, and rosmarinic acid were the metabolites responsible for these properties.

As mentioned before, some studies have evaluated the biological activity of plant extracts against phytopathogenic microorganisms *in vitro*; however, there are few works where it is evaluated *in vivo*, mainly in fruit postharvest storage. Xu et al. (2018) reported that phenolic compounds inhibited growth *in vitro* of *B. cinerea*, and treatment of grape berries with pterostilbene and piceatannol significantly reduced the disease incidence and severity.

Tanapichatsakul (2020) evaluated the volatile compounds of *Cuminum cyminum* on the mycelial growth of *Alternaria aculeatus* on grape berries and reported a reduced incidence and severity after ten days for storage at room temperature.

As we can see, these are just a few examples of the use of plant extracts for the control and management of plant-pathogenic microorganisms. Using plant extracts has several advantages in preventing the development of resistance, due to the usual presence of several bioactive compounds, low persistence in the environment, and the generally low cost of use.

7.6 SECONDARY METABOLITES WITH BIOLOGICAL ACTIVITY AGAINST PHYTOPATHOGENS: *IN VITRO* PRODUCTION

The *in vitro culture* of differentiated cells is one strategy for producing bioactive secondary metabolites (Choudhary et al. 2021). This alternative strategy eliminates the need to rely on wild-harvested plants (Espinosa-Leal et al. 2018).

A few studies have focused on the use of *in vitro* culture to produce secondary metabolites with biopesticide activity.

In this regard, one of the most studied systems *in vitro* is the culture of neem plants (*Azadirachta indica*) as the extracts have been used as a natural pesticide. Ashokhan et al. (2020) evaluated the effect of plant growth regulators (PGRs), thiadiazuron (TDZ) and 2,4-dichlorophenoxyacetic acid (2,4-D), on the induction of colored callus and the accumulation of azadirachtin (AZA). The authors reported that, through the use of PGRs, green-, brown-, or cream-colored callus were obtained, and that green callus extract contained the highest level of AZA (214.53 ± 33.63 mg/g DW) compared to brown- or cream-colored callus.

Another interesting research study involves the design, exploration, and analysis of a bioprocess plant that would produce a biopesticide from *A. indica* cell culture in a 10 m^3 bioreactor. The main botanical biopesticide compounds identified from neem culture were limonoids which act against almost 600 species of insects, mites, and nematodes (Martínez-Mira et al. 2020).

On the other hand, it has been reported that the aqueous extract and essential oil of the genus *Artemisia* have antibacterial, antifungal, repellent, insecticidal, nematicidal, and phytotoxic properties (Ivanescu et al. 2021). In the same way, Mohiuddin et al. (2018) designed an *in vitro* culture experiment to produce callus from *Artemisia absinthium*, using various PGRs, singly or in combination. The authors found that 0.75 mg/L NAA, 0.5 mg/L NAA, 0.5 mg/L 2,4-D + 0.5 mg/L NAA and 1 mg/L 2,4-D + 0.5 mg/L NAA induced the maximum callogenic response (100%) in the explants from leaves; in *Artemisia pallens*, of the growth regulators employed, the maximum callogenic response of 92% was observed at 2 mg/L kinetin from leaf explants. Phytochemical screening from the crude extracts showed the presence of glycosides, terpenoids, steroids, alkaloids, phenols, and tannins.

Koo et al. (2022) established *in vitro* culture conditions for the induction and proliferation of *Pinus strobus* callus from zygotic embryo cultures; various PGR concentrations were tested to assess the optimal concentration of 2,4-D. The authors reported that combined treatment with benzyladenine (BA) and 2,4-D stimulated the growth of callus (fresh weight) compared to 2,4-D alone. The accumulation of two pinosylvin stilbene compounds increased with the aging of the callus (after two or three months). Finally, the nematicidal activity of crude ethanolic extracts of callus were evaluated and only one of these compounds was effective for the immobilization of pine wood nematodes (PWNs: *Bursaphelenchus xylophilus*), reaching 67% immobilization after 24 h.

Thymus vulgaris and *Ocimum basilicum* are other species that have been investigated regarding the volatile constituents produced from cell suspensions and their role as insecticides against *Rhynchophorus ferrugineus*, which is a dangerous polyphagous insect, one of the most important pests of several palm species. In this research, the maximum cell suspension weights were obtained after 40 days. From the insecticidal test, the volatile extract of *T. vulgaris* was the most active, causing feeding inhibition against *R. ferrugineus* larvae (Darrag et al. 2021).

As can be seen, *in vitro* culture is a good alternative to wild-harvested plant tissue for producing metabolites of agricultural and pharmaceutical importance.

7.7 CONCLUSIONS AND FUTURE PROSPECTS

Throughout the world, food security, and the yield of cereals, fruits and vegetables are affected by fungi, bacteria, nematodes, insect pests, and viruses. The use of synthetic pesticidal chemical compounds for the control of plant phytopathogenic microorganisms has generated significant problems, including selection for pesticide resistance. Due to the above, environmentally friendly alternatives are necessary for management and control. Research evidence has shown that extracts obtained from plants protect against plant-pathogenic microorganisms and insect pests. However, these studies have been conducted at the laboratory level and plants were obtained from the wild. Therefore, it is necessary to research the optimal extraction method and its management on a commercial scale to achieve a sustainable approach. It is necessary to consider that the biosynthesis of metabolites is different at each stage of plant development and that environmental factors may also influence the process, with *in vitro* culture being an alternative.

REFERENCES

Abubakar Y., H. Tijjani, C. Egbuna, C. O. Adetunji, S. Kala, T. L. Kryeziu, J. C. Ifemeje, and K. C. Patrick-Iwuanyanwu. 2020. "Pesticides, history, and classification." In *Natural Remedies for Pest, Disease and Weed Control*, edited by C. Egbuna and B. Sawicka, 29–42. Academic Press. https://doi.org/10.1016/b978-0-12-819304-4.00003-8

Aguilar-Veloz L. M., M. Calderón-Santoyo, and J. A. Ragazzo-Sánchez. 2021. "Optimization of microwave assisted extraction of *Artocarpus heterophyllus* leaf polyphenols with inhibitory action against *Alternaria* sp. and antioxidant capacity." *Food Science and Biotechnology* 30(51): 1695–707. https://doi.org/10.1007/s10068-021-00996-8

Almeida F., M. L. Rodrigues, and C. Coelho. 2019. "The still underestimated problem of fungal diseases worldwide." *Frontiers Microbiology* 10: 214. http://doi.org/10.3389/fmicb.2019.00214

Al-Rahmah A. N., A. A. Mostafa, A. Abdel-Megeed, S. M. Yakout, and S. A. Hussein. 2013. "Fungicidal activities of certain methanolic plant extracts against tomato phytopathogenic fungi." *African Journal of Microbiology Research* 7(6): 517–24. https://doi.org/10.5897/AJMR12.1902

Ashokhan S., R. Othman, M. H. A. Rahim, S. A. Karsani, and J. S. Yaacob. 2020. "Effect of plant growth regulators on coloured callus formation and accumulation of azadirachtin, an assential biopesticide in *Azadirachta indica*." *Plants* 9: 352. https://doi.org/10.3390/plants9030352

Bardin M., and M. L. Gullino. 2020. "Fungal diseases." In *Integrated Pest and Disease Management in Greenhouse Crops*, edited by M. L. Gullino, R. Albajes, and P. C. Nicot, 55–100. Cham: Springer. https://doi.org/10.1007/978-3-030-22304-5_3

Bayar Y., and M. Yilar. 2019. "The antifungal and phytotoxic effect of different plant extracts of *Salvia virgata* Jacq." *Fresenius Environmental Bulletin* 28(4A): 3492–7. https://hdl.handle.net/20.500.12513/4000

Benelli G., R. Pavela, F. Maggi, J. G. N. Wandjou, N. B. Y. Fofie, D. Koné-Bamba, G. Sagratini, S. Vittori, and G. Caprioli. 2019. "Insecticidal activity of the essential oil and polar extracts from *Ocimum gratissimum* grown in Ivory Coast: Efficacy on insect pests and vectors and impact on non-target species." *Industrial Crops and Products* 132: 377–85. https://doi.org/10.1016/j.indcrop.2019.02.047

Bhattacharyya A., P. Duraisamy, M. Govindarajan, A. A. Buhroo, and R. Prasad. 2016. "Nano-biofungicides: Emerging trend in insect pest control." In *Advances and Applications Through Fungal Nanobiotechnology: Fungal Biology*, edited by R. Prasad, 307–19. Cham: Springer. https://doi.org/10.1007/978-3-319-42990-8_15

Catara V., and B. Patrizia. 2020. "Bacterial diseases in integrated pest and disease management." In *Greenhouse Crops*, edited by M. L. Gullino, R. Albajes, and P. Nicot, 33–54. Cham: Springer Nature. https://doi.org/10.1007/978-3-030-22304-5_2

Chandran H., M. Meena, T. Barupal, and K. Sharma. 2020. "Plant tissue culture as a perpetual source for production of industrially important bioactive compounds." *Biotechnology Reports* 26: e00450. https://doi.org/10.1016/j.btre.2020.e00450

Charkowski A. O. 2018. "The changing face of bacterial soft-rot diseases." *Annual Review of Phytopathology* 56: 269–88. https://doi.org/10.1146/annurev-phyto-080417-045906

Chen C., L. Long, F. Zhang, Q. Chen, C. Chen, X. Yu, Q. Liu, J. Bao, and Z. Long. 2018. "Antifungal activity, main active components and mechanism of *Curcuma longa* extract against *Fusarium graminearum*." *PLoS One* 13(3): e0194284. https://doi.org/10.1371/journal.pone.0194284

Chiocchio I., M. Mandrone, P. Tomasi, L. Marincich, and F. Poli. 2021. "Plant secondary metabolites: An opportunity for circular economy." *Molecules* 26(2): 495. https://doi.org/10.3390/molecules26020495

Choudhary P., R. Aggarwal, S. Rana, R. Nagarathnam, and M. Muthamilarasan. 2021. "Molecular and metabolomic interventions for identifying potential bioactive molecules to mitigate diseases and their impacts on crop plants." *Physiological and Molecular Plant Pathology* 114: 101624. https://doi.org/10.1016/j.pmpp.2021.101624

Cluzet S., J.-M. Mérillon, and K. G. Ramawat. 2020. "Specialized metabolites and plant defence." In *Plant Defence: Biological Control*, edited by J.-M. Mérillon and K. G. Ramawat, 45–80. Cham: Springer. https://doi.org/10.1007/978-3-030-51034-3_2

Cóndor G. A. F. 2019. "*In vitro* study on the nematicidal effect of different plant extracts on *Pratylenchus penetrans* and *Meloidogyne chitwoodi*." *Revista Facultad Nacional de Agronomía Medellín* 72(3): 8945–52. https://doi.org/10.15446/rfnam.v72n3.76070

Cotes B., B. Rämert, and U. Nilsson. 2018. "A first approach to pest management strategies using trap crops in organic carrot fields." *Crop Protection* 112: 141–8. https://doi.org/10.1016/j.cropro.2018.05.025

Daradka H. M., A. Saleem, and W. A. Obaid. 2021. "Antifungal effect of different plant extracts against phytopathogenic fungi *Alternaria alternata* and *Fusarium oxysporum* isolated from tomato plant." *Journal of Pharmaceutical Research International* 33(31A): 188–97. https://doi.org/10.9734/JPRI/2021/v33i31A31681

Darrag H. M., M. R. Alhajhoj, and H. E. Khalil. 2021. "Bio-insecticide of *Thymus vulgaris* and *Ocimum basilicum* extract from cell suspensions and their inhibitory effect against serine, cysteine, and metalloproteinases of the red palm weevil (*Rhynchophorus ferrugineus*)." *Insects* 12(5): 405. https://doi.org/10.3390/insects12050405

das Freire M. G. M., V. Mussi-Dias, T. C. Mattoso, D. A. Henk, A. Mendes, M. L. R. Macedo, C. Turatti, W. Machado, and I. Samuels. 2017. "Survey of endophytic Alternaria species isolated from plants in the Brazilian restinga biome." *IOSR Journal of Pharmacy and Biological Sciences* 12(2): 84–94. https://doi.org/10.9790/3008-1202038494

De Freitas S. M., V. P. Campos, A. F. Barros, J. C. P. da Silva, M. P. Pedroso, F. de J. Silva, V. A. Gomes, and J. C. Justino. 2020. "Medicinal plant volatiles applied against the root-knot nematode Meloidogyne incognita." *Crop Protection* 130: 105057. https://doi.org/10.1016/j.cropro.2019.105057

Djian-Caporalino C. 2012. Root-knot nematodes (*Meloidogyne* spp.), a growing problem in French vegetable crops. *Bulletin OEPP/EPPO Bulletin* 42(1): 127–37. https://doi.org/10.1111/j.1365-2338.2012.02530.x

Douglas A. E. 2018. "Strategies for enhanced crop resistance to insect pests." *Annual Review of Plant Biology* 69: 637–60. https://doi.org/10.1146/annurev-arplant-042817-04024

Dowling M., N. Peres, S. Villani, and G. Schnabel. 2020. "Managing *Colletotrichum* on fruits crops: A complex challenge." *Plant Disease* 104: 2301–16. https://doi.org/10.1094/PDIS-11-19-2378-FE

El-Hefny M., N. A. Ashmawy, M. Z. M. Salem, and A. Z. M. Salem. 2017. "Antibacterial activities of the phytochemicals-characterized extracts of *Callistemon viminalis*, *Eucalyptus camaldulensis* and *Conyza dioscoridis* against the growth of some phytopathogenic bacteria." *Microbial Pathogenesis* 113: 348–56. https://doi.org/10.1016/j.micpath.2017.11.004

Espinosa-Leal C. A., C. A. Puente-Rarza, and S. Garcia-Lara. 2018. "*In vitro* plant tissue culture means for production of biological active compounds." *Planta* 248: 1–18. https://doi.org/10.1007/s00425-018-2910-1

Evangelista-Lozano S., L. Reyes-Vaquero, A. de Jesús-Sánchez, S. V. Ávila-Reyes, A. R. Jiménez-Aparicio, and M. Y. Ríos. 2018. "Chemistry and insecticide activity of *Bougainvillea glabra* Choisy against *Spodoptera frugiperda* Smith." *Journal of Agriculture and Life Sciences* 5(2): 38–45. https://doi.org/10.30845/jals.v5n2p6

Figueiredo J., P. Vieira, I. Abrantes, and I. Esteves. 2021. "Detection of the root lesion nematode *Pratylenchus penetrans* in potato tubers." *Plant Pathology* 70(8): 1960–8. https://doi.org/10.1111/ppa.13425

Ghareeb R. Y., E. E. Hafez, and D. S. S. Ibrahim. 2020. "Current management strategies for phytoparasitic nematodes." In *Management of Phytonematodes: Recent Advances and Future Challenges*, edited by R. A. Ansari, R. Rizvi, and I. Mahmood, 339–52. Singapore: Springer. https://doi.org/10.1007/978-981-15-4087-5_15

He H.-M., L.-N. Liu, S. Munir, N. H. Bashir, Y. Wang, J. Yang, and C.-Y. Li. 2019. "Crop diversity and pest management in sustainable agriculture." *Journal of Integrative Agriculture* 18(9): 1945–52. https://doi.org/10.1016/S2095-3119(19)62689-4

Hussain T., A. Muhammad, K. Sarzamin, S. Hamid, and Q. M. Subhan. 2011. "*In vitro* screening of methanol plant extracts for their antibacterial activity." *Pakistan Journal of Botany* 43(1): 531–38.

International Plant Protection Convention (IPPC). 2021. "International year of plant health - Final report. Protecting plants, protecting life." FAO on behalf of the Secretariat of the International Plant Protection Convention. https://doi.org/10.4060/cb7056en

Ivanescu B., A. F. Burlec, F. Crivoi, C. Rosu, and A. Corciova. 2021. "Secondary metabolites from *Artemisia* genus as biopesticides and innovative nano-based application strategies." *Molecules* 26: 3061. https://doi.org/10.3390/molecules26103061

Jackson M., and B. M. Chen-Charpentier. 2018. "A model of biological control of plant virus propagation with delays." *Journal of Computational and Applied Mathematics* 330(1): 855–65. https://doi.org/10.1016/j.cam.2017.01.005

Jain A., S. Sarsaiya, Q. Wu, Y. Lu, and J. Shi. 2019. "A review of plant leaf fungal diseases and its environment speciation." *Bioengineered* 10(1): 409–24. https://doi.org/10.1080/21655979.2019.1649520

Jiang G., Z. Wei, J. Xu, H. Chen, Y. Zhang, X. She, A. P. Macho, W. Ding, and B. Liao. 2017. "Bacterial wilt in China: History, current status, and future perspectives." *Frontiers in Plant Science* 8: 1549. https://doi.org/10.3389/fpls.2017.01549

Karkanis A. C., and C. G. Athanassiou. 2020. "Natural insecticides from native plants of the Mediterranean basin and their activity for the control of major insect pests in vegetable crops: Shifting from the past to the future." *Journal of Pest Science* 94: 187–202. https://doi.org/10.1007/s10340-020-01275-x

Kaur S., S. Pandey, and S. Goel. 2019. "Plants disease identification and classification through leaf images: A survey." *Archives of Computational Methods in Engineering* 26: 507–30. https://doi.org/10.1007/s11831-018-9255-6

Khan S., C. N. T. Taning, E. Bonneure, S. Mangelinckx, G. Smagghe, and M. M. Shah. 2017. "Insecticidal activity of plant-derived extracts against different economically important pest insects." *Phytoparasitica* 45: 113–24. https://doi.org/10.1007/s12600-017-0569-y

Khare S., N. B. Singh, A. Singh, I. Hussain, K. Niharika, V. Yadav, C. Bano, R. K. Yadav, and N. Amist. 2020. "Plant secondary metabolites synthesis and their regulations under biotic and abiotic constraints." *Journal of Plant Biology* 63: 203–16. https://doi.org/10.1007/s12374-020-09245-7

Khumbuzile N. B., Y. Petersen, C. T. Bull, and T. A. Coutinho. 2020. "Identification of *Pseudomonas* isolates associated with bacterial canker of stone fruit trees in the Western Cape, South Africa." *Plant Disease* 104: 882–92. https://doi.org/10.1094/PDIS-05-19-1102-RE

Koo H. B., H.-S. Hwang, J. Y. Han, E. J. Cheong, Y.-S. Kwon, and Y. E. Choi. 2022. "Enhanced production of pinosylvin stilbene with aging of *Pinus strobus* callus and nematicidal activity of callus extracts against pinewood nematodes." *Scientific Reports* 12: 770. https://doi.org/10.1038/s41598-022-04843-6

Kumar V., and N. Iqbal. 2020. "Postharvest pathogens and disease management of horticultural crop: A brief review." *Plant Archives* 20(2): 2054–58.

Lefebvre T., E. Destandau, and E. Lesellier. 2021. "Selective extraction of bioactive compounds from plants using recent extraction techniques: A review." *Journal of Chromatography A* 635: 461770. https://doi.org/10.1016/j.chroma.2020.461770

Lengai G. M. W., J. W. Muthomi, and E. R. Mbega. 2020. "Phytochemical activity and role of botanical pesticides in pest management for sustainable agricultural crop production." *Scientific African* 7: e00239. https://doi.org/10.1016/j.sciaf.2019.e00239

Luna-Guevara J. J., M. M. P. Arenas-Hernandez, C. M. Peña, J. L. Silva, and M. L. Luna-Guevara. 2019. "The role of pathogenic *E. coli* in fresh vegetables: Behavior, contamination factors, and preventive measures." *International Journal of Microbiology* 2019: 2894328. https://doi.org/10.1155/2019/2894328

Mandal H. R., S. Katel, S. Subedi, and J. Shrestha. 2021. "Plant parasitic nematodes and their management in crop production: A review." *Journal of Agriculture and Natural Resources* 4(2): 327–38. https://doi.org/10.3126/janr.v4i2.33950

Marrone P. G. 2019. "Pesticidal natural products - Status and future potential." *Pest Management Science* 75(9): 2325–40. https://doi.org/10.1002/ps.5433

Martínez-Mira A., A. Vásquez-Rivera, R. Hoyos-Sánchez, and F. Orozco-Sánchez. 2020. "Bioprocess plant design and economic analysis of an environmentally friendly insect controller agent produced with *Azadirachta indica* cell culture." *Biochemical Engineering Journal* 159: 107579. https://doi.org/10.1016/j.bej.2020.107579

Mendez M., R. Rodríguez, J. Ruiz, D. Morales-Adame, F. Castillo, F. D. Hernández-Castillo, and C. N. Aguilar. 2012. "Antibacterial activity of plant extracts obtained with alternative organics solvents against food-borne pathogen bacteria." *Industrial Crops and Products* 37(1): 445–50. https://doi.org/10.1016/j.indcrop.2011.07.017

Mesa-Valle C. M., J. A. Garrido-Cardenas, J. Cebrian-Carmona, M. Talavera, and F. Manzano-Agugliaro. 2020. "Global research on plant nematodes." *Agronomy* 10(8): 1148. https://doi.org/10.3390/agronomy10081148

Mohiuddin Y. G., V. N. Nathar, W. N. Aziz, and N. Gaikwad. 2018. "Induction of callus and preliminary phytochemical profiling from callus of *Artemisia absinthium* L. and *Artemisia pallens* Wall." *International Journal of Current Trends in Science and Technology* 8(2): 20236–41. http://currentsciences.info/10.15520/ctst.v8i02.285.pdf

Mori Y., Y. Kuwano, S. Tomokiyo, N. Kuroyanagi, and K. Odahara. 2019. "Inhibitory effects of moso bamboo (*Phyllostachys heterocycla* f. pubescens) extracts on phytopathogenic bacterial and fungal growth." *Wood Science and Technology* 53: 135–50. https://doi.org/10.1007/s00226-018-1063-5

Moriones E., and E. Verdin. 2020. "Viral diseases." In *Integrated Pest and Disease Management in Greenhouse Crops*, edited by M. L. Gullino, R. Albajes, and P. C. Nicot, 3–31. Cham: Springer. https://doi.org/10.1007/978-3-030-22304-5_1

Pham D. Q., D. T. Ba, N. T. Dao, G. J. Choi, T. T. Vu, J.-C. Kim, T. P. L. Giang, H. D. Vu, and Q. L. Dang. 2017. "Antimicrobial efficacy of extracts and constituents fractionated from *Rheum tanguticum* Maxim. ex Balf. rhizomes against phytopathogenic fungi and bacteria." *Industrial Crops and Products* 108: 442–50. http://doi.org/10.1016/j.indcrop.2017.06.067

Poveda J., D. Eugui, and P. Velasco. 2020. "Natural control of plant pathogens through glucosinolates: An effective strategy against fungi and oomycetes." *Phytochemistry Reviews* 19: 1045–59. https://doi.org/10.1007/s11101-020-09699-0

Pradhan A., S. Ghosh, D. Sahoo, and G. Jha. 2020. "Fungal effectors, the double edge sword of phytopathogens." *Current Genetics* 67: 27–40. https://doi.org/10.1007/s00294-020-01118-3

Rashid U., A. Panhwar, A. Farhan, M. Akhter, N. Jalbani, and D. R. Hashmi. 2021. "Nematicidal effects of various fractions of *Curcuma longa* against *Meloidogyne incognita* (root knot nematodes)." *Turkish Journal of Agricultural Engineering Research* 2(1): 175–82. https://doi.org/10.46592/turkager.2021.v02i01.013

Raveau R., J. Fontaine, and A. L. H. Sahraoui. 2020. "Essential oils as potential alternative biocontrol products against plant pathogens and weeds: A review." *Foods* 9(3): 365. https://doi.org/10.3390/foods9030365

Reyes-Vaquero L., M. Bueno, R. I. Ventura-Aguilar, A. B. Aguilar-Guadarrama, N. Robledo, G. Sepúlveda-Jiménez, P. E. Vanegas-Espinoza, E. Ibáñez, and A. A. Del Villar-Martínez. 2021. "Seasonal variation of chemical profile of *Ruta graveolens* extracts and biological activity against *Fusarium oxysporum, Fusarium proliferatum and Stemphylium vesicarium.*" *Biochemical Systematics and Ecology* 95: 104223. https://doi.org/10.1016/j.bse.2021.104223

Reyes-Vaquero L., M. E. Valdés-Estrada, A. R. Jiménez-Aparicio, S. L. Escobar-Arellano, and S. Evangelista-Lozano. 2016. "Evaluation of methanolic extract of *Bougainvillea glabra* Choisy "Variegata" against *Spodoptera frugiperda* under laboratory conditions." *Southwestern Entomologist* 41(4): 983–90. http://doi.org/10.3958/059.041.0428

Shaheena H. A., and M. Y. Isss. 2020. "*In vitro* and *in vivo* activity of *Peganum harmala* L. alkaloids against phytopathogenic bacteria." *Scientia Horticulturae* 264: 108940. https://doi.org/10.1016/j.scienta.2019.108940R

Shin-Ichi A. 2014. "*Pectobacterium carotovorum* - Subpolar Hyper-Flagellation." *The Flagellar World.* https://doi.org/10.1016/B978-0-12-417234-0.00018-9

Shuping D. S. S., and J. N. Eloff. 2017. "The use of plants to protect plants and food against fungal pathogens: A review." *African Journal of Traditional, Complementary, and Alternative Medicines* 14(4): 120–27. https://doi.org/10.21010/ajtcam.v14i4.14

Sithole N. T., M. G. Kulkarni, J. F. Finnie, and J. Van Staden. 2021. "Potential nematicidal properties of plant extracts against *Meloidogyne incognita.*" *South African Journal of Botany* 139: 409–17. https://doi.org/10.1016/j.sajb.2021.02.014

Sivanandhan S., P. Ganesan, A. Jackson, S. Darvin, M. G. Paulraj, and S. Ignacimuthu. 2018. "Activity of some medicinal plants against phytopathogenic fungi." *International Journal of Scientific Research in Biological Sciences* 5(5): 124–37. https://doi.org/10.26438/ijsrbs/v5i5.124137

Sledz W., E. Los, A. Paczek, J. Rischka, A. Motyka, S. Zoledowska, J. Piosik, and E. Lojkowska. 2015. "Antibacterial activity of caffeine against plant pathogenic bacteria." *Acta Biochimica Polonica* 62(3): 605–12. https://doi.org/10.18388/abp.2015_1092

Smant G., J. Helder, and A. Goverse. 2018. "Parallel adaptations and common host cell responses enabling feeding of obligate and facultative plant parasitic nematodes." *The Plant Journal* 93(4): 686–702. https://doi.org/10.1111/tpj.13811

Tajmiri P., S. A. A. Fathi, A. Golizadeh, and G. Nouri-Ganbalani. 2017. "Effect of strip-intercropping potato and annual alfalfa on populations of *Leptinotarsa decemlineata* Say and its predators." *International Journal of Pest Management* 63(4): 273–9. https://doi.org/10.1080/09670874.2016.1256513

Tanapichatsakul C., S. Khruengsai, and P. Pripdeevech. 2020. "*In vitro* and *in vivo* antifungal activity of *Cuminum cyminum* essential oil against *Aspergillus aculeatus* causing bunch rot of postharvest grapes." *PLoS One* 15(11): e0242862. https://doi.org/10.1371/journal.pone.0242862

Tawfeeq A. A. L. K., and E. L. Furtado. 2020. "The effect of incompatible plant pathogens on the host plant." In *Molecular Aspects of Plant Beneficial Microbes in Agriculture*, edited by V. Sharma, R. Salwan, and L. K. T. Al-Ani, 47–57. Academic Press. https://doi.org/10.1016/B978-0-12-818469-1.00004-3

Tembo Y., A. G. Mkindi, P. A. Mkenda, N. Mpumi, R. Mwanauta, P. C. Stevenson, P. A. Ndakidemi, and S. R. Belmain. 2018. "Pesticidal plant extracts improve yield and reduce insect pests on legume crops without harming beneficial arthropods." *Frontiers in Plant Science* 9: 1425. https://doi.org/10.3389/fpls.2018.01425

Vázquez-González Y., J. A. Ragazzo-Sánchez, and M. Calderón-Santoyo. 2020. "Characterization and antifungal activity of jackfruit (*Artocarpus heterophyllus* Lam.) leaf extract obtained using conventional and emerging technologies." *Food Chemistry* 330: 127211. https://doi.org/10.1016/j.foodchem.2020.127211

Vázquez-Sánchez M., J. R. Medina-Medrano, H. Cortez-Madrigal, M. V. Angoa-Pérez, C. V. Muñoz-Ruíz, and E. Villar-Luna. 2018. "Nematicidal activity of wild plant extracts against second-stage juveniles of *Nacobbus aberrans*." *Nematropica* 48(2): 136–44.

Velasco-Azorsa R., H. Cruz-Santiago, I. C. del Prado-Vera, M. V. Ramírez-Mares, M. del R. Gutiérrez-Ortiz, N. F. Santos-Sánchez, R. Salas-Coronado, C. Villanueva-Cañongo, K. I. Lira-de León, and B. Hernández-Carlos. 2021. "Chemical characterization of plant extracts and evaluation of their nematicidal and phytotoxic potential." *Molecules* 26(8): 2216. https://doi.org/10.3390/molecules26082216

Wang K., S. Jiang, T. Pu, L. Fan, F. Su, and M. Ye. 2019. "Antifungal activity of phenolic monoterpenes and structure-related compounds against plant pathogenic fungi." *Natural Product Research* 33(10): 1423–30. https://doi.org/10.1080/14786419.2017.1419232

Xu D., Y. Deng, T. Han, L. Jiang, P. Xi, Q. Wang, Z. Jiang, and L. Gao. 2018. "*In vitro* and *in vivo* effectiveness of phenolic compounds for the control of postharvest gray mold of table grapes." *Postharvest Biology and Technology* 139: 106–14. https://doi.org/10.1016/j.postharvbio.2017.08.019

Yao S., J. Xu, H. Zhang, H. Gao, S. Shang, and D. Wang. 2021. "*In vitro* activity of extracts of five medicinal plant species on plant pathogenic fungi." *Frontiers of Agricultural Science and Engineering* 8(4): 635–44. https://doi.org/10.15302/J-FASE-2020343

Zaynab M., M. Fatima, S. Abbas, Y. Sharif, M. Umair, M. H. Zafar, and K. Bahadar. 2018. "Role of secondary metabolites in plant defense against pathogens." *Microbial Pathogenesis* 124: 198–202. https://doi.org/10.1016/j.micpath.2018.08.034

Zivanovic M., and R. R. Walcott. 2017. "Further characterization of genetically distinct groups of *Acidovorax citrulli* strains." *Phytopathology* 107(1): 29–35. https://doi.org/10.1094/PHYTO-06-16-0245-R

CHAPTER 8

Encapsulation of Plant Secondary Metabolites of Industrial Interest

Jorge A. Ramos-Hernández,
Carla Norma Cruz-Salas,
Katia Nayely González-Gutiérrez,
Elda Margarita González-Cruz,
and Cristina Prieto López

CONTENTS

DOI: 10.1201/9781003166535-8

8.1 INTRODUCTION

Plants have been used as a source of compounds with healing properties since ancient times (Harvey et al. 2015). In this sense, scientific research has demonstrated that these healing properties are mainly due to plant secondary metabolites, which have been demonstrated to exhibit bioactive properties (Augustin and Sanguansri 2012). However, these compounds have low stability and are susceptible to degradation (Dima et al. 2015). Since these compounds exhibit a wide range of natural biological functions, significant activity against pathogens, and with minimal or no side effects, it is of great importance to ensure the protection and improve the bioavailability of these metabolites. Therefore, the development of encapsulation systems is an important strategy for the protection, controlled release, and delivery of the compound. In addition, their encapsulation facilitates their handling and storage (Cruz-Salas et al. 2019).

In recent years, the bioactive properties of secondary metabolites have attracted considerable interest from the pharmaceutical and food industry, which are promoting the study of new sources, novel extraction, purification, and encapsulation processes, as well as alternative administration routes for the benefit of human health. The objective of this chapter is to present the innovations in extraction and protection processes for secondary metabolites and their importance in the food and health industry.

8.2 OVERVIEW OF SECONDARY METABOLITES

Secondary metabolites can be defined as organic compounds produced by an organism, which do not directly intervene in their metabolism, but are used as a defense mechanism, or to attract pollinators (Ahmed et al. 2017). Most of the secondary metabolites are of plant origin; however, some of them are of microbial or animal origin, such as the ω-3 fatty acids from fish (Santana-Gálvez et al. 2019), hydrolyzed meat peptides (Pogorzelska-Nowicka et al. 2018), and indolocarbazole from the blue shrimp, *Litopenaeus stylirostris* (García-Romo et al. 2020), among others.

8.2.1 Plant Sources of Secondary Metabolites

Secondary metabolites of plant origin are mainly found in leaves, stems, flowers, roots, barks, seeds, or fruits. In this regard, metabolites in plants come from photosynthetic metabolic routes, such as the shikimate pathway of phenylpropanoids synthesis and the mevalonate pathway of terpenoid (e.g., squalene) synthesis (Azmir et al. 2013; Embuscado 2015).

According to Singh and Shukla (2015), secondary metabolites are chemically classified as terpenoids, alkaloids, steroids, phenolic compounds, carbohydrate derivatives, fatty acids/lipids, and amino acid derivatives. These compounds have been demonstrated to have great potential and a positive influence on human health.

8.3 IMPORTANCE OF SECONDARY METABOLITES IN HUMAN HEALTH

Regarding the use of secondary metabolites in the biomedical area, these phytochemicals can be considered compounds able to strengthen the immune system, inhibit viral replication processes, or carry out signaling processes to activate the defense mechanisms against diseases. Since these compounds exhibit a wide range of natural biological functions, and minimal or no side effects, they are interesting candidates for pharmaceutical research (Gandhi et al. 2015; Gorlenko et al. 2020; Li et al. 2020).

Some of the beneficial properties reported for phenolic compounds are related to the reduction of the risk of chronic degenerative diseases, the prevention of cardiovascular diseases and diabetes, and the improvement of visual acuity due to the antioxidant, antimicrobial, anticancer, anti-inflammatory, and chemoprotective properties of these compounds (Xu et al. 2017). These properties are related to the high electron-donating capacity of the hydroxyl groups present in these molecules. However, the beneficial effects of polyphenols may be due not only to their antioxidant capacity, but also to their impact on enzymes, cell signaling pathways, and effects on gene expression (Wangensteen et al. 2015). Furthermore, polyphenols, together with flavonoids, coumarins, terpenoids, and other substances, represent the antiplatelet effect. These compounds correct platelet abnormalities by interfering with different signaling pathways, including inhibition of the adenosine diphosphate (ADP) pathway, suppression of thromboxane (TXA_2) formation, reduction of the intracellular mobilization of Ca^{2+}, and degradation of phosphoinositides (Haouari and Rosado 2016).

Alkaloids are another interesting group of active metabolites since they possess potential therapeutic effects against several neurodegenerative diseases. This group of bioactive compounds includes neuroactive molecules, such as caffeine and nicotine as well as important active pharmaceutical ingredients, including emetine, used to combat oral poisoning, and the antitumor drugs vincristine and vinblastine (Matsuura and Fett-Neto 2015). As well, this group of bioactive compounds includes other molecules such as isoquinoline, indole, pyrroloindole, oxindole, piperidine, pyridine, aporphine, vinca, β-carboline, methylxanthine, and erythrine.

Alkaloids function as substances that bind to a receptor on a cell, known as muscarinic or adenosine receptor agonists, but also as antioxidants, anti-amyloids, monoamine oxidase inhibitors, acetylcholinesterase and butyrylcholinesterase inhibitors, and dopaminergic and nicotine agonists (Hussain et al. 2018). The action of alkaloids and amines often occurs on neuroreceptors, modulating the transduction of neuronal signals, including ion channels or enzymes, which capture or metabolize neurotransmitters or second messengers. Also, it has been reported that some alkaloids interfere with the biosynthesis of DNA, telomeres, telomerase, topoisomerase, cytoskeleton, or proteins, and they can induce apoptosis (Wink 2015).

Terpenes constitute another important group, which, as a result of their biological properties, have been used by the pharmaceutical and biotechnology industries to guide the synthesis of drugs (Habib et al. 2020). Within this chemical group, pentacyclic triterpenes have gained attention due to their anti-inflammatory, antibacterial, antioxidant, antitumor, anti-HIV, hepatoprotective, and immuno-adjuvant properties (Xu et al. 2018). The presence of pentacyclic triterpenes, such as lupeol and α-and β-amyrin, has been reported in various plant species; however, *Coccoloba uvifera,* also known as sea grape, stands out among the species studied for its high content of these compounds (Ramos-Hernández et al. 2018a). The use of α and β-amyrin has been indicated in the preventive or therapeutic treatment of colitis. Moreover, α-and β-amyrin decrease colon pro-inflammatory mediators, tumor necrosis factor (TNF) -α, interleukin (IL) -1β, and keratinocyte-derived chemokine (CXCL1/KC), and they also regulate IL-4 levels (Matos et al. 2013). Another compound of terpenic origin is lupeol acetate, which is associated with the control of the symptoms of rheumatoid arthritis by regulating the expressions of inflammatory cytokines and osteoclastogenesis (Wang et al. 2016). Oleanolic and ursolic acids are compounds that have been evaluated for ophthalmic administration with favorable corneal permeation and release results (Alvarado et al. 2015). Another reported example is the successful use of the terpene-rich extract of the leaves of *Lantana camara* for the development of drugs against *Mycobacterium tuberculosis* (Patil and Kumbhar 2018). This extract presents a high concentration of β-caryophyllene that can be used to suppress the growth of *M. tuberculosis* (H37Ra).

With regard to carotenoids, compounds such as β-carotene, lutein, lycopene, and cryptoxanthin have been reported for their antimutagenic and antiproliferative potential (Ruiz-Montañez et al. 2015; Rivera-Aguilar et al. 2020). Plant species, such as jackfruit (*Artocarpus heterophyllus*), have been demonstrated to be rich sources of these compounds, which could be considered for use in the treatment of diseases associated with oxidative stress (Ruiz-Montañez et al. 2015). In addition, jackfruit contains metabolites with important antioxidant, antimutagenic, and antiproliferative activities observed during ripening and consumption (Kumar et al. 2021; Ruiz-Montañez et al. 2019). Rivera-Aguilar et al. (2020) and Hosen et al. (2022) reported the presence in jackfruit of compounds such as apigenin, artocarpine, quadraflavone C, moracin C, and β-carotene. Some of these compounds have shown biological activity against the carcinogenesis process at different phases.

Another important factor in the treatment of some chronic diseases is the control of inflammatory processes. In this sense, the nuclear factors kappaB (NF-κB) and STAT3 have emerged as important regulators of inflammation, cell transformation, and survival, proliferation, invasion, angiogenesis, and metastasis of tumor cells. Thus, secondary metabolites that can inhibit the activation pathways of NF-κB and STAT3 have the potential to prevent and treat cancer. For instance, triterpenes have been considered as interesting compounds for their ability to suppress inflammatory pathways related to tumorigenesis (Cháirez-Ramírez et al. 2015; Liu et al. 2021).

However, despite the interesting bioactive properties shown by these compounds, their potential health benefits depend on the absorption and metabolism which, in turn, are determined by the structure, molecular size, solubility, and stability of these molecules. In this sense, the extraction, purification, and encapsulation of these compounds will help to maintain their bioactive properties.

8.4 SECONDARY METABOLITES ISOLATION TECHNIQUES

Solvent extraction, known as maceration, is the most widely used method for the recovery of secondary metabolites from a wide range of matrices of plant origin. The objective of this process is to separate compounds from a solid matrix, based on the principle of solubility. The solvent diffuses into the matrix and dissolves the related compounds (Strati and Oreopoulou 2014). The polarity of the target metabolite is the most important factor in the choice of solvent since the success of this operation depends on the molecular affinity between the solvent and the solute (Ramos-Hernández et al. 2021). The main solvents used are water, methanol, ethanol, acetone, ethyl acetate, chloroform, hexane, and petroleum ether, as well as mixtures of polar and non-polar solvents in different ratios. The main parameters to be considered in an extraction process are the kind of solvent, type of sample, extraction temperature, and extraction time (Ramos-Hernández et al. 2018a; Vázquez-González et al. 2020).

The main disadvantages of this method are the heavy consumption of organic solvents, the long extraction periods, and the low yields of extracted metabolites (Ramos-Hernández et al. 2018a). In order to improve the extraction yields, it is necessary to review and evaluate non-conventional, environmentally safe techniques for the recovery of metabolites. In this context, the most reported techniques are ultrasound-assisted extraction (UAE), microwave-assisted extraction (MAE), high hydrostatic pressure extraction (HHPE), enzyme-assisted extraction (EAE), and supercritical fluid-assisted extraction (SFAE).

UAE is mainly based on acoustic cavitation caused by the interaction between the ultrasonic waves, the liquid, and the dissolved gas. This phenomenon breaks the cell walls and favors the release of compounds. These waves, at certain frequencies and amplitudes, create cavitation bubbles which, when they reach a non-stable point, release high temperature and high pressure by imploding (Lefebvre, Destandau, and Lesellier 2021). UAE is an effective technique for the extraction of plant secondary metabolites. Ramos-Hernández et al. (2018a) showed that UAE is more effective than maceration in obtaining lupeol from leaves of *C. uvifera*, reporting 5.61 mg/g d.b. (dried base) compared to 1.11 mg/g d.b. obtained by maceration.

MAE uses microwave technology in order to heat a solid and liquid mixture. Heating is induced by microwave irradiation absorbed by molecules depending on their dielectric constant. Polar molecules with a high dielectric constant are able to absorb this energy and re-emit it, causing the heating of the system. Therefore, the energy absorption by the solvent mixture increases, leading to an easier release of compounds (Lefebvre et al. 2021). Lerma-Torres et al. (2019) reported that MAE allows a higher concentration of hexane extract (0.25 mg/g d.b.) to be achieved from *Mangifera indica* leaves, including mangiferin and lupeol, compared with maceration or heat (< 0.2 mg/g d.b.).

HHPE is a non-thermal technology that has been identified as being environmentally friendly. HHPE is conducted in the pressure range from

100 to 800 MPa. The extraction stage is usually applied through a three-stage process: pressurizing stage, pressure maintenance stage, and pressure relief stage (Giacometti et al. 2018). In the first stage, the disruption of plant tissue, the cell wall and the membrane enhances the mass transfer of the soluble constituents to the solvent. During the second stage, the pressure inside and outside the cells is balanced, so that solvent permeates the cells rapidly and contacts the cellular contents, dissolving the active component in a shorter period with respect to other conventional techniques. Finally, in the third stage, the pressure is relieved in a few seconds, during which time pressure within cells rapidly decreases with the alteration of the non-covalent bonds. This causes cells to expand, resulting in subsequent increases in permeability and decreased resistance to diffusion and permeation. After pressure release, the resulting mixture should be separated and purified (Ferrentino et al. 2018; Giacometti et al. 2018). Vázquez-González et al. (2020) obtained phenols (> 155 mg gallic acid-equivalents, GAE/g d.b.) from the jackfruit leaf by HHPE at 200 and 150 MPa for 15 min, whereas less than 40 mg GAE/g d.b. were obtained using conventional extraction methods for 24 h.

EAE is an unconventional technique based on the hydrolysis of the cell walls by specific enzymes, such as cellulases, pectinases, and hemicellulases, allowing the efficient release and extraction of secondary metabolites within the cells. Enzymes hydrolyze the cell wall components and alter the structural integrity of the plant cell wall. Consequently, they increase the permeability of the solvent to the cell wall, and thus, higher extraction yields of SM are achieved (Tomaz et al. 2016). Catalkaya and Kahveci (2019) obtained lycopene from tomato (11.5 mg/g) and conclude that EAE is an effective technique.

Finally, supercritical fluids represent a new generation of green solvents. In particular, supercritical CO_2 is non-toxic, non-flammable, relatively cheap, has mild critical conditions ($T_C = 304$ K and $P_C = 7.4$ MPa), and can be recovered and recycled after use (Prieto et al. 2017). The process takes place at reduced temperature in the absence of light, and it is easy to separate the solvent and extracts. In addition, specific fractions can be collected by altering the extraction parameters, such as pressure and temperature (Giacometti et al. 2018). In this way, SFAE is based on the modulation of these fluids' properties (density, diffusivity, dielectric constant, and viscosity) with pressure and temperature variations, allowing the extraction and recovery of the bioactive components of the sample. Zekovic et al. (2015) used SFAE to extract essential oils (EO) from sweet basil (*Ocimum basilicum*), obtaining an extraction yield of 0.56 % (v/w), with the major compounds being linalool (10.14 to 49.79%, w/w), eugenol (3.74 to 9.78%) and δ-cardinene (3.94 to 8.07%).

8.5 ENCAPSULATION OF SECONDARY METABOLITES TO PROTECT THEIR BIOLOGICAL ACTIVITY

The secondary metabolites obtained from advanced extraction techniques are of great interest, but they have low stability and are susceptible to decomposition caused by environmental factors such as oxygen, ultraviolet light, temperature, and humidity (Ramos-Hernández et al. 2020). In addition, most of the secondary metabolites, when administered orally, can decrease or lose their functional activity due to factors such as the acidic pH of the stomach, the alkaline pH of the intestine, or enzymatic activity (Braithwaite et al. 2014).

Likewise, a variation in the physical and/or chemical properties of the extracted secondary metabolites affects their bioactive properties and their bioavailability. In this sense, hydrophobic metabolites, such as curcumin, have low solubility in water, so the bioavailability is also low, which also limits the biological activity of these molecules (Zhang et al. 2020). Similarly, pentacyclic triterpenes such as lupeol, α-, and β-amyrin have a hydrophobic structure (Ramos-Hernández et al. 2018a), limiting their solubility in water. Some secondary metabolites such as polyphenols are unstable and are easily oxidized during storage (González-Cruz et al. 2020). Similarly, β-carotene is an unstable molecule in response to ultraviolet light and oxygen, factors that cause the decrease or loss of its biological activity (Ramos-Hernández et al. 2018b). Other compounds, such as saponins, have a poor taste and texture, which limit their role by oral administration (Dawid and Hofmann 2014).

To produce effective responses, metabolites must maintain their bioactive properties and they must be successfully delivered to tissue or organs at a safe and effective dose level. Therefore, it is of great importance to ensure their oxidative stability and increase their water solubility and pH resistance (Zhang et al. 2020). In this sense, the administration systems based on encapsulation have exhibited excellent yields in the protection, controlled release, and targeted delivery of secondary metabolites to the specific site (Kotta et al. 2012).

During the encapsulation process, the bioactive compound is trapped inside an inert matrix, which isolates and protects it from the external environment. Moreover, encapsulation can also serve to achieve controlled release, mask undesired organoleptic properties, prevent adverse effects, increase shelf-life, or provide easier handling (Weiss et al. 2006). The effectiveness of the encapsulation process will depend on the type of encapsulating matrix selected and the encapsulation process chosen.

Materials commonly used as encapsulation matrices are polysaccharides, such as chitosan, alginates, celluloses, starches, or malto- and cyclodextrins. In addition, natural hydrocolloids, such as whey protein isolate (WPI), whey protein concentrate (WPC), collagen, and gelatin (Echegoyen et al. 2017) have also been used for encapsulation purposes. Biopolymers are preferably used because they are biocompatible, biodegradable, and have a high potential for modification to achieve the required properties (Esfanjani and Jafari 2016). Furthermore, encapsulating materials can have significant stabilizing effects on the texture of foods (Alehosseini et al. 2018). The most promising technologies for encapsulating SMs are nanoemulsions, spray drying, and electrohydrodynamic processes (electrospraying, electrospinning, and coaxial electrohydrodynamic process).

8.5.1 Encapsulation of Secondary Metabolites by Nanoemulsion

Emulsions can be defined as a colloidal dispersion, made up of two immiscible liquids, which is obtained by dispersing very fine drops of one liquid (called the dispersed or internal phase) within another one (called the continuous or external phase) (Akhavan et al. 2018). Nanoemulsions are defined as stable ultrafine dispersions, composed of two immiscible liquids, either oil in water (O/W) or water in oil (W/O), with a mean droplet diameter of 20 to 500 nm (Singh et al. 2017; Karami et al. 2019; Sharma et al. 2019). Nanoemulsions are made up of at least three essential components: oil phase, aqueous phase, and surfactants/co-surfactants (Karami et al. 2019). The oils commonly used in the formulation of nanoemulsions are sesame oil, peanut oil, corn oil, lanolin, jojoba oil, Capryol 90, isopropyl palmitate, palm oil esters, miglyol, castor oil, olive oil, soybean oil, oleic acid, pine nut oil, or coconut oil, among others (Shah et al. 2017; Singh et al. 2017). To maintain stability and reduce interfacial tension, surfactants such as hydrophilic colloids, Span 80, Tween 20, Tween 80, powdered soy lecithin, sodium caseinate, egg lecithin, and sulfosuccinate have been used so far (Sharma et al. 2019).

Nanoemulsions are made primarily by two processes, including the high-energy emulsification method (input energy 10^8–10^{10} W/kg) and the low-energy emulsification method (input energy ~ 10^3–10^5 W/kg) (Sharma et al. 2019). In high-energy methods, large disruptive forces are provided through the use of mechanical devices, including ultrasonic emulsification, high-pressure homogenization, high-energy agitation, and microfluidization, to break the macroemulsion droplets into smaller droplets (Shah et al. 2017). Low-energy methods use the intrinsic physiological properties of the system for the production of nanoemulsions. These methods include phase inversion temperature, emulsion inversion point, solvent displacement, and self-emulsification (Aswathanarayan and Vittal 2019; Karamiet al. 2019; Jiang et al. 2020).

Some examples are cinnamon oil nanoemulsions, vitamin E and vitamin D acetate nanoemulsions, fish oil, and cinnamaldehyde nanoemulsions (Aswathanarayan and Vittal 2019).

Nanoemulsions are widely regarded as one of the most popular delivery systems for a wide range of secondary metabolites with a lipophilic, hydrophilic, and amphiphilic nature (Lu et al. 2016). Oil-in-water nanoemulsions have been used to deliver vitamins, flavors, antioxidants, preservatives, or nutraceuticals (Acevedo-Fani et al. 2017), as summarized in Table 8.1.

In other studies, phenolics, essential oils, and flavonoids administered as nanoemulsions have exhibited biological activities greater than the pure free molecules; examples are the nanoemulsion of curcumin composed of medium-chain triglycerides and phospholipids, which proved to be efficient as a photosensitizing agent with great potential for using in breast cancer treatment (Machado et al. 2019). The oregano oil nanoemulsion, made with sunflower oil, castor oil, and Span 80 by the phase inversion temperature method, was suitable for incorporation into food formulations to prevent and control microbial growth (Moraes-Lovison et al. 2017). In addition, the luteolin nanoemulsion, made with lipoid lecithin and almond oil using phase inversion and sonication methods, promoted hair growth at a level similar to that of commercially available minoxidil (Shin et al. 2018).

In the pharmaceutical industry, nanoemulsions can be used in liquids, creams, aerosols, gels, or foams, and can be administered by different routes, such as topical, oral, intravenous, intranasal, pulmonary, and ocular (Singh et al. 2017). In this way, nanoemulsions trapped in structured organogels have been developed to increase the oral bioavailability of curcumin. In addition, nanoemulsions have been explored for the topical uptake of drugs; they provide a combination of penetration enhancement (by altering lipid bilayers) and a concentration gradient by acting as tiny reservoirs of drugs. W/O nanoemulsions have been

TABLE 8.1 Examples of Recent Studies of Encapsulated Secondary Metabolites with Application in the Food and Pharmaceutical Industries

Secondary Metabolite (Source)	Encapsulation Material	Encapsulation Method	System	Application in the Industry	Reference
Curcumin (*Curcuma longa* L./ dried rhizomes)	Medium-chain-triglycerides, natural soy phospholipids Anionic surfactant, poloxamer 188	Spontaneous nanoemulsification	Nanoemulsion	It proved to be efficient as a photosensitizing agent, had phototoxic effects, decreased the proliferation of MCF-7 cells and stimulated the production of reactive oxygen species, with great potential for use in the treatment of breast cancer.	(Machado et al. 2019)
Oregano oil (*Origanum vulgare* L./oregano plant)	Sunflower oil, PEG-40 hydroxylated castor oil, sorbitan monooleate (Span 80).	Phase inversion temperature		It was suitable to be incorporated into food formulations, such as chicken pâté, to prevent and control microbial growth.	(Moraes-Lovison et al. 2017)
Luteolin (*Dracocephalum ruyschiana* L./ leaves)	Lipoid lecithin P75-3, sweet almond oil.	Phase inversion composition, followed by probe-type sonication.		It promoted hair growth at a level similar to that of the commercially available minoxidil.	(Shin et al. 2018)

(*Continued*)

TABLE 8.1 (CONTINUED) Examples of Recent Studies of Encapsulated Secondary Metabolites with Application in the Food and Pharmaceutical Industries

Secondary Metabolite (Source)	Encapsulation Material	Encapsulation Method	System	Application in the Industry	Reference
Anthocyanins (*Solanum tuberosum* L./purple-fleshed cultivated potato)	Maltodextrin	Spray drying	Microcapsules	They showed stability during storage and within the *in vitro* gastrointestinal digestion model. Furthermore, microcapsules may have potential use as colorants or powder formulations for the food and health industry.	(Vergara et al. 2020)
Phenolic compounds (*Mangifera indica* L./seeds)	Maltodextrin	Spray drying		Encapsulated phenolic compounds showed properties for application in the pharmaceutical and food industry.	(Siacor et al. 2020)
Thyme oil (*Thymus vulgaris* L./thyme plant)	Sodium casein and maltodextrin	Spray drying		Encapsulated thyme essential oil showed antioxidant and antimicrobial activity against *Staphylococcus aureus, Escherichia coli, Listeria monocytogenes*, and *Salmonella typhimurium* tested *in vitro* and against thermotolerants coliforms and *Escherichia coli in situ*. The potential for its application as a natural food preservative was evidenced.	(Radünz et al. 2020a)

(*Continued*)

TABLE 8.1 (CONTINUED) Examples of Recent Studies of Encapsulated Secondary Metabolites with Application in the Food and Pharmaceutical Industries

Secondary Metabolite (Source)	Encapsulation Material	Encapsulation Method	System	Application in the Industry	Reference
Glucosinolates and phenolic compounds (*Brassica oleracea* L. var. *italica*/broccoli florets)	Zein	Electrospraying	Capsules	The encapsulated extract showed strong selective antitumor effect against glial tumor cells without toxicity to non-tumor cells.	(Radünz et al. 2020b)
Polyphenols (*Euterpe oleracea* Mart/açaí fruit)	Zein	Electrospraying		They improvement of thermal stability of açaí fruit polyphenols after sterilization (121 °C) and baking (180 °C). Bioaccessibility studies indicated an increase in the presence of polyphenols after the *in vitro* digestion steps.	(Dicastillo et al. 2019)
Phenols (*Olea europea* cv. Mission/olive leaf)	WPC	Electrospraying		They can be applied for biofunctional food production and as nanodelivery systems in pharmaceuticals.	(Soleimanifar, Jafari, and Assadpour 2020)
Polyphenolic antioxidants (*Momordica charantia* L./bitter melon fruit)	Zein, gelatin.	Coaxial electrospinning	Fibers	They showed potential to be used as stand-alone nutraceutical supplement products or as an ingredient (filling or edible wrapper) in food products.	(Torkamani et al. 2018)

(*Continued*)

TABLE 8.1 (CONTINUED) Examples of Recent Studies of Encapsulated Secondary Metabolites with Application in the Food and Pharmaceutical Industries

Secondary Metabolite (Source)	Encapsulation Material	Encapsulation Method	System	Application in the Industry	Reference
Carvacrol (*Thymus vulgaris* L./ thyme plant)	Soluble potato starch	Electrospinning		The structures are promising materials for application as a carvacrol release vehicle in antimicrobial and antioxidant food packages.	(Fonseca et al. 2019)
Thyme essential oil (*Thymus vulgaris* L./ thyme plant)	Gelatin	Electrospinning		They were a promising prospect in the preservation of chicken meat without negative impact on sensory properties.	(Lin, Zhu, and Cui 2018)

used to deliver proteins or plasmids via the transepidermal route. Their oil component is compatible with sebum present in follicular openings which act as entry points. This opens up the possibility of directing and confining potentially therapeutic transgenics to clinically active skin lesions. On the other hand, ocular administration of O/W nanoemulsions has been attempted to deliver water-incompatible, environmentally sensitive, poorly absorbed (via the trans-corneal route), or poorly retained drugs (Singh et al. 2017).

However, the stability of nanoemulsions depends on the storage temperature, pH, and ionic strength. To obtain a stable nanoemulsion and prevent coalescence, cremation, flocculation, and sedimentation, the optimization of preparation conditions and the composition of nanoemulsions are required (Singh et al. 2017; Sharma et al. 2019). On the other hand, they present advantages, such as the delivery of metabolites to a specific site and the improvement of absorption through cells. The lipids used are biocompatible, biodegradable, and non-mutagenic, so nanoemulsions are safe for human health. On the other hand, there is a perception in the food industry that, in the manufacture of nanoemulsions, the use of high concentrations of emulsifiers makes them expensive products, as well as the fact that they are labile at low pH values, while the controlled release process does not take place due to their small size and the liquid nature of the vehicle. Nevertheless, the pharmaceutical and food industries have already launched nanoemulsions on the market and, with the arrival of new instruments and competition between leading companies, the cost of preparing the nanoemulsion will decline and will probably reach the price of regular macroemulsions (Akhavan et al. 2018).

8.5.2 Encapsulation of Secondary Metabolites by Spray Drying

Spray drying is a well-established microencapsulation technique commonly used in the food and pharmaceutical industries. The spray-drying process is inexpensive and can preserve the natural properties by trapping the active ingredient within a coating (Ruiz-Montañez et al. 2015), which can be a polysaccharide, such as gum arabic, maltodextrin, modified starch, sodium alginate, guar gum, chitosan, inulin, or proteins such as whey protein, soy protein, or sodium caseinate (Drosou et al., 212017; Flores and Kong 2017). Spray drying results in powders of good quality, with low water activity, which are easy to handle and store, and with protective effects on bioactive materials. Additionally, spray drying is a continuous, simple, and fast process (Ruiz-Montañez et al. 2015).

This process is based on the principle of dissolving or dispersing the active ingredients in a biopolymer solution (Bor et al. 2016). The dispersion is atomized in a drying chamber with a nozzle that makes the droplets as small as possible (Muhamad et al. 2017). These droplets come into contact with the hot drying air and their large surface area facilitates the heat transfer from the drying air to the atomized particles, leading to the rapid evaporation of the solvent that is available on the surface of the droplet (Ho et al. 2017). During this step, the mass of the droplet decreases while the temperature remains constant at the wet-bulb temperature of the solvent (Ziaee et al. 2019). A solid outer crust forms on the surface of the droplet, the concentration of ingredients in the drying droplet increases, while the smaller water molecules diffuse outwards until generating a dry particle that consists of the active ingredient embedded in the wall material. Finally, the particles fall to the bottom of the drying chamber, being separated from the hot air in the cyclone, where they are collected (Ho et al. 2017; Ziaee et al. 2019). Final powder properties can be altered by varying the formulation and process parameters (Ziaee et al. 2019), such as core and wall material properties, solution viscosity, and spray-drying parameters, such as flow, inlet/outlet temperature, and humidity (Talegaonkar et al. 2016).

Spray drying has been used in the dehydration of products to preserve them and facilitate their transport and storage (Ho et al. 2017; Jimenez-Sánchez et al. 2020). Recently, this technology has been used in the encapsulation of secondary metabolites (Table 8.1), including anthocyanins, mango fruit phenolic compounds, and essential oils. Maltodextrin was used as a wall material for the encapsulation of anthocyanins, and the obtained microcapsules showed a potential use as colorants or powdered ingredients for the food and health industry (Vergara et al. 2020). In another study, reported by Siacor et al. (2020), an extract of phenolic compounds from mango seed was encapsulated by spray drying. The total polyphenol content of the powders ranged between 50.62 and 96.14 mg of GAE per g of dry powder, with potential use in pharmaceutical and food applications. Spray drying has also been used in the encapsulation of essential oils, such as thyme oil. Radünz et al. (2020a) reported capsules with high encapsulation efficiency, excellent thermal stability, and regular morphology, with high antioxidant and antimicrobial activity, and the potential to act as a preservative in hamburgers (Table 8.1). In addition, some microorganisms such as *Lactobacillus acidophilus* (Arepally et al. 2020) and *Meyerozyma caribbica* (Aguirre-Güitrón et al. 2018), proteins (casein, immunoglobulin, bovine serum albumin) (Sarabandi et al. 2020), caffeine, vitamins (Flores and Kong 2017), minerals, dyes, enzymes, and oleoresins (Gupta et al. 2016) have been encapsulated. Among the encapsulating materials available, gum arabic, maltodextrin, modified starch, and zein have been reported as being used for the encapsulation of oleoresins, avoiding volatile losses and increasing their useful life (Drosou, Krokida, and Biliaderis 2017).

The dairy and food processing industries were the first to industrially use spray-drying technology, followed thereafter by the pharmaceutical and ceramic industries. Nowadays, spray drying is applied in the biopharmaceutical industry and it has been used for engineering specific particle sizes for pulmonary delivery applications e.g., insulin for the treatment of diabetes and vaccines with excipients (Ziaee et al. 2019). In the food industry, this encapsulation technique is highly focused on the encapsulation of flavors, spices, and colorants.

Spray drying is one of the most widely applied technologies because it is a low-cost, reproducible method of continuous large-scale production, easy to operate, and scalable (Aguirre-Güitrón et al. 2018). It provides relative control on the particle size distribution and high encapsulation efficiencies (Jimenez-Sánchez et al. 2020) at the same time as helping to preserve natural properties.

However, this process has some drawbacks. Aromas can be partially lost and oxidized during the drying process due to the high processing temperatures. In addition, some bioactive components can be surface exposed in the microparticles; for this reason, it is also considered an immobilization method (Gupta et al. 2016). Likewise, there may be other drawbacks, such as reduced yield due to the deposition of powder on the wall of the drying chamber because of the problem of stickiness, low efficiency of the separation cyclone, and the difficulty of obtaining stable, free-flowing amorphous products with coating materials exhibiting low glass transition temperatures (Ho et al. 2017).

8.5.3 Encapsulation of Secondary Metabolites by Electrohydrodynamic Processes

Electrohydrodynamic processes allow a polymer solution to be spun or pulverized by applying a high-potential electric field. Electrospinning and electrospraying are aimed at obtaining nanofibers and nanospheres, respectively. The equipment configuration consists of a high-voltage source

(1–30 kV), a stainless-steel injector, a syringe pump, and a grounded collector, which can be a flat plate or a rotating drum (Wen et al. 2017).

The principle of the electrospray technique is based on the fact that a liquid flowing from a capillary nozzle is subjected to electrical forces due to a strong electric field generated near the nozzle. Driven by the electric field, the electric charges move within the liquid and are distributed in such a way that the positive charge is enriched on the surface of the liquid located at the capillary tip, whereas the negative charge is directed towards the interior of the capillary (Alehosseini et al. 2018). The electric charge generates an electrostatic force that competes with the surface tension of the droplet, forming the Taylor cone. This cone results from an effect of the balance between surface tension and electrical and gravitational forces. When the electrical charge accumulated at the vertex of the cone exceeds the surface tension of the liquid, a jet of charged liquid is expelled from the tip of the Taylor cone and fragmented into droplets due to the electrical repulsion of the charges placed on its surface (Esfanjani and Jafari 2016). Subsequently, the excess charge must be dissipated, which causes the explosion of the primary droplet, generating smaller droplets on the micro- or nanoscale. Due to the repulsion of charges, the droplets disperse and do not agglomerate on their way to the collector, during which the solvent evaporates from the charged droplets. Finally, the solid polymeric particles settle down in the collector (Esfanjani and Jafari 2016).

In electrospinning, the main electrostatic forces, such as the electrostatic repulsion of similar charges and the Coulomb repulsion of the external electric field, cause the hemispherical surface of the droplet to distort, forming the Taylor cone (Drosou et al. 2017; Wen et al. 2017). As the voltage increases, the surface tension and the viscoelastic forces balance, allowing the formation of a straight and stable jet (Lasprilla-Botero et al. 2018). As the jet is directed towards the collector, unevenly distributed electrical charges cause bending movements in the jet, resulting in the elongation of the jet and favoring the rapid evaporation of the solvent. Subsequently, an ultrafine and solid fiber is deposited on the collector (Wen et al. 2017).

The spheres and fibers obtained by these processes have structural and functional advantages. They can have adaptable size and morphology, structures on a micro-, submicron-, or nanoscale, a high ratio of surface area to volume, a highly porous structure (Jafari 2017; Wen et al. 2017), as well as provide controlled release of the encapsulated material and high encapsulation efficiency (Ramos-Hernández et al. 2020). Furthermore, the secondary metabolites have greater stability and functionality since they do not require severe conditions of pressure, aggressive chemicals, or temperature (Echegoyen et al. 2017; Wen et al. 2017). In electrospraying, the use of food-grade biopolymers guarantees its application in a food or pharmaceutical product (Cruz-Salas et al. 2019). Moreover, there is a high level of control over the particle size distribution, with the ability to produce quasi-monodisperse particles. However, more studies are required to expand their range of applications in the food and pharmaceutical industries (Drosou, Krokida, and Biliaderis 2017). The electrospinning process depends at the same time on a considerable number of interrelated variables, so defining the set of conditions that work for specific polymer/solvent systems sometimes requires investing time in optimizing the process (Lasprilla-Botero et al. 2018).

Based on these advantages, electrohydrodynamic processes show a high application potential. Electrospraying has been used for the stabilization of nutraceuticals, production of pharmaceutical particles, improvement of the viability of probiotic substances, delivery systems for controlled release, and micro-nano-encapsulation of secondary metabolites (Alehosseini et al. 2018). Electrospinning, on the other hand, has been used to encapsulate bioactive compounds, prepare nano-biosensors, immobilize enzymes, controlled release delivery systems, for preparation of edible films and packaging materials, and the production of nanostructured layers for food packaging (Wen et al. 2017).

Regarding micro-nano-encapsulation of secondary metabolites, previous research has revealed that, through electrohydrodynamic processes, the stability and bioavailability (Fonseca et al. 2019; Soleimanifar et al. 2020), as well as the thermal stability (Dicastillo et al. 2019) of the metabolites can be improved to be used as independent nutraceutical complementary products or as ingredients in food products (Torkamani et al. 2018).

Electrohydrodynamic technology is highly versatile in terms of encapsulating materials. Previous research has reported the use of proteins, carbohydrates, or synthetic polymers. The encapsulating materials based on proteins which are commonly used in these processes are whey protein isolate (WPI), amaranth, zein (Alehosseini et al. 2018), whey protein concentrate (WPC), soy protein isolate (Basar et al. 2020), as well as egg albumen, collagen, gelatin, and casein. Carbohydrate-based biopolymers include ß-cyclodextrin, chitosan, pullulan (Drosou et al. 2017; Wen et al. 2017), and, recently, high-polymerization agave fructans have also been included (Ramos-Hernández et al. 2018b; Cruz-Salas et al. 2019). Furthermore, other polymers, such as polyacrylonitrile, polyethylene oxide, polyvinyl alcohol, and poly (3-hydroxybutyrate), have been used (Lasprilla-Botero et al. 2018). These biopolymers showed a clear and strong potential to develop spheres and fibers with high encapsulation yield and excellent loading capacity to improve the delivery of secondary metabolites (Alehosseini et al. 2018).

Some examples of secondary metabolites encapsulated by electrospraying with application in the food and pharmaceutical industries are shown in Table 8.1. Glucosinolates and phenolic compounds were encapsulated in zein, presenting a selective antitumor effect against glial tumor cells (Radünz et al. 2020b). Another study demonstrated that the thermal stability was improved in zein-encapsulated açaí fruit polyphenols after *in vitro* digestion stages (Dicastillo et al. 2019). In a recent study, phenols of the olive leaf were encapsulated by electrospraying with WPC, showing that the nanoparticles have the potential for use in biofunctional foods and as nanodelivery systems in pharmaceutical products (Soleimanifar et al. 2020). Some examples of secondary metabolites encapsulated by electrospinning are presented in Table 8.1. Starch nanofibers encapsulating carvacrol are promising materials for their use as carvacrol release vehicles in food packaging (Fonseca et al. 2019). On the other hand, thyme essential oil was encapsulated within gelatin nanofibers, and it was considered to be a promising prospect for the preservation of chicken meat without affecting the sensory properties (Lin et al. 2018).

In general, the main disadvantage of the electrohydrodynamic processes is their low productivity, normally with a processing throughput of a few milliliters per hour per single emitter, which has limited their widespread use for industrial applications (Prieto and Lagaron 2020). In this context, companies like Bioinicia S. L. (Paterna, Spain) have developed electrohydrodynamic plants for manufacturing fibers or particles on an industrial scale. As well, Bioinicia, together with the research group of Lagaron, have developed an innovative encapsulation technique based on the combination of electrospraying with the pneumatic atomization process. This novel high-throughput technology, termed electrospraying assisted by pressurized gas (EAPG), is based on the atomization of the polymer solution by a pneumatic injector using compressed air that nebulizes within a high electric field. During this process, the solvent is evaporated at room temperature in an evaporation chamber and the encapsulated material is then collected as a free-flowing powder in a cyclone (Prieto and Lagaron 2020). By means of this technology, it is possible to reach the production yields required by the pharmaceutical and food industries. The potential of this novel technology has recently been proven for the encapsulation of omega-3 oils (Prieto and Lagaron 2020).

8.5.3.1 Encapsulation of Secondary Metabolites by Coaxial Electrospray

This process involves the use of two concentrically arranged needles connected to two tanks containing different solutions, one with the polymeric matrix and the other with the core material. This strategy serves to overcome the disadvantage of component immiscibility (Drosou et al. 2017; Echegoyen et al. 2017).

The coaxial method allows the production of core-shell particles or fibers containing the secondary metabolite in the inner injector and the wall material in the outer injector (Puppi and Chiellini 2018). Advantages of coaxial electrospraying include high encapsulation efficiency, effective protection of the properties/functionality of the encapsulated metabolite, as well as a uniform particle size distribution (Rathore and Schiffman 2021). Moreover, to ensure that the secondary metabolites have minimal contact with the organic solvent of the polymer solution, the three-needle coaxial devices can allow more types of drugs to be loaded into separate capsule layers for sequential and multiple releases (Gómez-Mascaraque et al. 2019). This approach can facilitate the production of multi-axial nanofiber designs that exhibit three (or more) distinct layers on individual fibers, so that compounds can be loaded into any or all of the layers to produce systems with highly specific and effective uses (Wen et al. 2017).

In this regard, the potential of coaxial processes to produce multilayer encapsulation structures has been demonstrated in food, pharmaceutical, and biomedical applications (Gómez-Mascaraque et al. 2019). These processes have been used for the encapsulation of liquid compounds with self-healing applications (Cuvellier et al. 2018) and in the encapsulation of growth factors, DNA, living organisms, enzymes, and secondary metabolites (Puppi and Chiellini 2018). As shown in Table 8.1, one example of the encapsulation of these metabolites by coaxial electrospinning is polyphenolic antioxidants, in which the fibers showed potential uses as nutraceutical complementary products or as an edible filler or wrap ingredient in food products (Torkamani et al. 2018).

8.6 CONCLUSIONS

Plant secondary metabolites possess a wide range of functional properties related to human health. However, they also represent an important challenge in the development of new technologies, as well as how to take advantage of their mechanisms of action for food and human health benefit with the aim of being used more efficiently.

Extraction and isolation depend mainly on the methods used and their subsequent applications. In this sense, operating parameters must be optimized to maximize extraction performance. Among the available techniques, ultrasound-assisted extraction has demonstrated high efficiency and high yields in short process times compared with conventional techniques. Regarding encapsulation, electrohydrodynamic processes stand out because of their high encapsulation efficiency, flexibility, and their ability to better conserve metabolites sensitive to heat and oxidation.

However, more research is still needed to deepen our knowledge of the mechanisms of action of secondary metabolites, to obtain enhanced plant extracts with improved bioactive properties, and to obtain superior encapsulates with enhanced stability and bioavailability.

REFERENCES

Acevedo-Fani A., R. Soliva-Fortuny, and O. Martín-Belloso. 2017. "Nanoemulsions as edible coatings." *Current Opinion in Food Science* 15: 43–9. https://doi.org/10.1016/j.cofs.2017.06.002

Aguirre-Güitrón L., M. Calderón-Santoyo, R. I. Ortiz-Basurto, P. U. Bautista-Rosales, and J. A. Ragazzo-Sánchez. 2018. "Optimisation of the spray drying process of formulating the post-harvest biocontrol agent *Meyerozyma caribbica*." *Biocontrol Science and Technology* 28(6): 574–90. https://doi.org/10.1080/09583157.2018.1468997

Ahmed E., M. Arshad, M. Z. Khan, M. S. Amjad, H. M. Sadaf, I. Riaz, P. S. Sabir, N. Ahmad. 2017. "Secondary metabolites and their multidimensional prospective in plant life." *Journal of Pharmacognosy and Phytochemistry* 6(2): 205–14.

Akhavan S., E. Assadpour, I. Katouzian, and S. M. Jafari. 2018. "Lipid nano scale cargos for the protection and delivery of food bioactive ingredients and nutraceuticals." *Trends in Food Science and Technology* 74: 132–46. https://doi.org/10.1016/j.tifs.2018.02.001

Alehosseini A., B. Ghorani, M. Sarabi-Jamab, and N. Tucker. 2018. "Principles of electrospraying: A new approach in protection of bioactive compounds in foods." *Critical Reviews in Food Science and Nutrition* 58(14): 2346–63. https://doi.org/10.1080/10408398.2017.1323723

Alvarado H. L., G. Abrego, M. L. Garduño-Ramirez, B. Clares, A. C. Calpena, and M. L. García. 2015. "Design and optimization of oleanolic/ursolic acid-loaded nanoplatforms for ocular anti-inflammatory applications." *Nanomedicine: Nanotechnology, Biology, and Medicine* 11(3): 521–30. https://doi.org/10.1016/j.nano.2015.01.004

Arepally D., R. S. Reddy, and T. K. Goswami. 2020. "Encapsulation of *Lactobacillus acidophilus* NCDC 016 cells by spray drying: Characterization, survival after *in vitro* digestion, and storage stability." *Food and Function* 11(10): 8694–706. https://doi.org/10.1039/d0fo01394c

Aswathanarayan J. B., and R. R.Vittal. 2019. "Nanoemulsions and their potential applications in food industry." *Frontiers in Sustainable Food Systems* 3: 1–21. https://doi.org/10.3389/fsufs.2019.00095

Augustin M. A., and L. Sanguansri. 2012. "Challenges in developing delivery systems for food additives, nutraceuticals and dietary supplements." In *Encapsulation Technologies and Delivery Systems for Food Ingredients and Nutraceuticals*, edited by N. Garti and D. J. McClements, 19–48. Sawston: Woodhead Publishing. https://doi.org/10.1533/9780857095909.1.19

Azmir J., I. S. M. Zaidul, M. M. Rahman, K. M. Sharif, A. Mohamed, F. Sahena, M. H. A. Jahurul, K. Ghafoor, N. A. N. Norulaini, and A. K. M. Omar. 2013. "Techniques for extraction of bioactive compounds from plant materials: A review." *Journal of Food Engineering* 117(4): 426–36. https://doi.org/10.1016/j.jfoodeng.2013.01.014

Basar A. O., C. Prieto, E. Durand, P. Villeneuve, H. T. Sasmazel, and J. Lagaron. 2020. "Encapsulation of β-carotene by emulsion electrospraying using deep eutectic solvents." *Molecules* 25(4): 981. https://doi.org/10.3390/molecules25040981

Bor T., S. O. Aljaloud, R. Gyawali, and S. A. Ibrahim. 2016. "Antimicrobials from herbs, spices, and plants." In *Fruits, Vegetables, and Herbs: Bioactive Foods in Health Promotion*, edited by R. R. Watson and V. R. Preedy, 551–78. Cambridge: Academic Press. https://doi.org/10.1016/B978-0-12-802972-5.00026-3

Braithwaite M. C., C. Tyagi, L. K. Tomar, P. Kumar, Y. E. Choonara, and V. Pillay. 2014. "Nutraceutical-based therapeutics and formulation strategies augmenting their efficiency to complement modern medicine: An overview." *Journal of Functional Foods* 6(1): 82–99. https://doi.org/10.1016/j.jff.2013.09.022

Catalkaya G., and D. Kahveci. 2019. "Optimization of enzyme assisted extraction of lycopene from industrial tomato waste." *Separation and Purification Technology* 219: 55–63. https://doi.org/10.1016/j.seppur.2019.03.006

Cháirez-Ramírez M. H., J. A. Sánchez-Burgos, C. Gomes, M. R. Moreno-Jiménez, R. F. González-Laredo, M. J. Bernad-Bernad, L. Medina-Torres, M. V. Ramírez-Mares, J. A. Gallegos-Infante, and N. E. Rocha-Guzmán. 2015. "Morphological and release characterization of nanoparticles formulated with poly (DL-lactide-co-glycolide) (PLGA) and lupeol: *In vitro* permeability and modulator effect on NF-κB in Caco-2 cell system stimulated with TNF-α." *Food and Chemical Toxicology* 85: 2–9. https://doi.org/10.1016/j.fct.2015.08.003

Cruz-Salas C. N., C. Prieto, M. Calderón-Santoyo, J. M. Lagarón, and J. A. Ragazzo-Sánchez. 2019. "Micro-and nanostructures of *Agave* fructans to stabilize compounds of high biological value via electrohydrodynamic processing." *Nanomaterials* 9(12): 1–12. https://doi.org/10.3390/nano9121659

Cuvellier A., A. Torre-Muruzabal, N. Kizildag, L. Daelemans, Y. Ba, K. De Clerck, and H. Rahier. 2018. "Coaxial electrospinning of epoxy and amine monomers in a pullulan shell for self-healing nanovascular systems." *Polymer Testing* 69: 146–56. https://doi.org/10.1016/j.polymertesting.2018.05.023

Dawid C., and T. Hofmann. 2014. "Quantitation and bitter taste contribution of saponins in fresh and cooked white asparagus (*Asparagus officinalis* L.)." *Food Chemistry* 145: 427–36. https://doi.org/10.1016/j.foodchem.2013.08.057

Dicastillo C. L., C. Piña, L. Garrido, C. Arancibia, and M. J. Galotto. 2019. "Enhancing thermal stability and bioaccesibility of açaí fruit polyphenols through electrohydrodynamic encapsulation into zein electrosprayed particles." *Antioxidants* 8(10): 464. https://doi.org/10.3390/antiox8100464

Dima Ş., C. Dima, and G. Iordăchescu. 2015. "Encapsulation of functional lipophilic food and drug biocomponents." *Food Engineering Reviews* 7(4): 417–38. https://doi.org/10.1007/s12393-015-9115-1

Drosou C. G., M. K. Krokida, and C. G. Biliaderis. 2017. "Encapsulation of bioactive compounds through electrospinning/electrospraying and spray drying: A comparative assessment of food-related applications." *Drying Technology* 35(2): 139–62. https://doi.org/10.1080/07373937.2016.1162797

Echegoyen Y., M. J. Fabra, J. L. Castro-Mayorga, A. Cherpinski, and J. M. Lagaron. 2017. "High throughput electro-hydrodynamic processing in food encapsulation and food packaging applications: Viewpoint." *Trends in Food Science and Technology* 60: 71–9. https://doi.org/10.1016/j.tifs.2016.10.019

Embuscado M. E. 2015. "Spices and herbs: Natural sources of antioxidants - A mini review." *Journal of Functional Foods* 18: 811–9. https://doi.org/10.1016/j.jff.2015.03.005

Esfanjani A. F., and S. M. Jafari. 2016. "Biopolymer nano-particles and natural nano-carriers for nano-encapsulation of phenolic compounds." *Colloids and Surfaces B: Biointerfaces* 146: 532–43. https://doi.org/10.1016/j.colsurfb.2016.06.053

Ferrentino G., M. Asaduzzaman, and M. M. Scampicchio. 2018. "Current technologies and new insights for the recovery of high valuable compounds from fruits by-products." *Critical Reviews in Food Science and Nutrition* 58(3): 386–404. https://doi.org/10.1080/10408398.2016.1180589

Flores F. P., and F. Kong. 2017. "*In vitro* release kinetics of microencapsulated materials and the effect of the food matrix." *Annual Review of Food Science and Technology* 8: 237–59. https://doi.org/10.1146/annurev-food-030216-025720

Fonseca L. M., C. E. S. Cruxen, G. P. Bruni, Â. M. Fiorentini, E. R. Zavareze, L. T. Lim, and A. R. G. Dias. 2019. "Development of antimicrobial and antioxidant electrospun soluble potato starch nanofibers loaded with carvacrol." *International Journal of Biological Macromolecules* 139: 1182–90. https://doi.org/10.1016/j.ijbiomac.2019.08.096

Gandhi S. G., V. Mahajan, and Y. S. Bedi. 2015. "Changing trends in biotechnology of secondary metabolism in medicinal and aromatic plants." *Planta* 241(2): 303–17. https://doi.org/10.1007/s00425-014-2232-x

García-Romo J. S., L. Noguera-Artiaga, A. C. Gálvez-Iriqui, M. S. Hernández-Zazueta, D. F. Valenzuela-Cota, R. I. González-Vega, M. Plascencia-Jatomea, M. G. Burboa-Zazueta, E. Sandoval-Petris, R. M. Robles-Sánchez, J. Juárez, J. Hernández-Martínez, H. C. Santacruz-Ortega, and A. Burgos-Hernández. 2020. "Antioxidant, antihemolysis, and retinoprotective potentials of bioactive lipidic compounds from wild shrimp (*Litopenaeus stylirostris*) muscle." *CYTA - Journal of Food* 18(1): 153–63. https://doi.org/10.1080/19476337.2020.1719210

Giacometti J., D. B. Kovačević, P. Putnik, D. Gabrić, T. Bilušić, G. Krešić, V. Stulić, F. J. Barba, F. Chemat, G. Barbosa-Cánovas, and A. R. Jambrak. 2018. "Extraction of bioactive compounds and essential oils from mediterranean herbs by conventional and green innovative techniques: A review." *Food Research International* 113: 245–62. https://doi.org/10.1016/j.foodres.2018.06.036

Gómez-Mascaraque L. G., F. Tordera, M. J. Fabra, M. Martínez-Sanz, and A. Lopez-Rubio. 2019. "Coaxial electrospraying of biopolymers as a strategy to improve protection of bioactive food ingredients." *Innovative Food Science and Emerging Technologies* 51: 2–11. https://doi.org/10.1016/j.ifset.2018.03.023

González-Cruz E. M., M. Calderón-Santoyo, J. C. Barros-Castillo, and J. A. Ragazzo-Sánchez. 2020. "Evaluation of biopolymers in the encapsulation by electrospraying of polyphenolic compounds extracted from blueberry (*Vaccinium corymbosum* L.) variety Biloxi." *Polymer Bulletin* 78: 3561–76. https://doi.org/10.1007/s00289-020-03292-3

Gorlenko C. L., H. Y. Kiselev, E. V. Budanova, A. A. Zamyatnin, and L. N. Ikryannikova. 2020. "Plant secondary metabolites in the battle of drugs and drug-resistant bacteria: New heroes or worse clones of antibiotics?" *Antibiotics* 9(4): 170. https://doi.org/10.3390/antibiotics9040170

Gupta S., S. Khan, M. Muzafar, M. Kushwaha, A. K. Yadav, and A. P. Gupta. 2016. "Encapsulation: Entrapping essential oil/flavors/aromas in food." In *Encapsulations*, edited by A. M. Grumezescu, 229–68. Cambridge: Academic Press. https://doi.org/10.1016/b978-0-12-804307-3.00006-5

Habib F., D. A. Tocher, N. J. Press, and C. J. Carmalt. 2020. "Structure determination of terpenes by the crystalline sponge method." *Microporous and Mesoporous Materials* 308: 110548. https://doi.org/10.1016/j.micromeso.2020.110548

Haouari M. E., and J. A. Rosado. 2016. "Medicinal plants with antiplatelet activity." *Phytotherapy Research* 1071: 1059–71. https://doi.org/10.1002/ptr.5619

Harvey A. L, R. Edrada-Ebel, and R. J. Quinn. 2015. "The re-emergence of natural products for drug discovery in the genomics era." *Nature Reviews Drug Discovery* 14: 111–29. https://doi.org/10.1038/nrd4510

Ho T. M, T. Truong, and B. Bhandari. 2017. "Spray-drying and non-equilibrium states/glass transition." In *Non-Equilibrium States and Glass Transitions in Foods*, edited by B. Bhandari and Y. H. Roos, 111–36. Sawston: Woodhead Publishing. https://doi.org/10.1016/B978-0-08-100309-1/00008-0

Hosen S. M. Z., M. Junaid, M. S. Alam, M. Rubayed, R. Dash, R. Akter, T. Sharmin, N. J. Mouri, M. A. Moni, M. Khatun, and M. Mostafa. 2022. "GreenMolBD: Nature derived bioactive molecules' database." *Medicinal Chemistry* 18(6): 724–33. https://doi.org/10.2174/1573406418666211129103458

Hussain G., A. Rasul, H. Anwar, N. Aziz, A. Razzaq, W. Wei, M. Ali, J. Li, and X. Li. 2018. "Role of plant derived alkaloids and their mechanism in neurodegenerative disorders." *International Journal of Biological Sciences* 14(3): 341–57. https://doi.org/10.7150/ijbs.23247

Jafari S. M. 2017. "An overview of nanoencapsulation techniques and their classification." In *Nanoencapsulation Technologies for the Food and Nutraceutical Industries*, 1–34. Cambridge: Academic Press. https://doi.org/10.1016/B978-0-12-809436-5.00001-X

Jiang T., W. Liao, and C. Charcosset. 2020. "Recent advances in encapsulation of curcumin in nanoemulsions: A review of encapsulation technologies, bioaccessibility and applications." *Food Research International* 132: 109035. https://doi.org/10.1016/j.foodres.2020.109035

Jimenez-Sánchez D. E., M. Calderón-Santoyo, E. Herman-Lara, C. Gaston-Peña, G. Luna-Solano, and J. A. Ragazzo-Sánchez. 2020. "Use of native agave fructans as stabilizers on physicochemical properties of spray-dried pineapple juice." *Drying Technology* 38(3): 293–303. https://doi.org/10.1080/07373937.2019.1565575

Karami Z., M. R. S. Zanjani, and M. Hamidi. 2019. "Nanoemulsions in CNS drug delivery: Recent developments, impacts and challenges." *Drug Discovery Today* 24(5): 1104–15. https://doi.org/10.1016/j.drudis.2019.03.021

Kotta S., A. W. Khan, K. Pramod, S. H. Ansari, R. K. Sharma, and J. Ali. 2012. "Exploring oral nanoemulsions for bioavailability enhancement of poorly water-soluble drugs." *Expert Opinion on Drug Delivery* 9(5): 585–98. https://doi.org/10.1517/17425247.2012.668523

Kumar M., J. Potkule, M. Tomar, S. Punia, S. Singh, S. Patil, S. Singh, T. Ilakiya, C. Kaur, and J. F. Kennedy. 2021. "Jackfruit seed slimy sheath, a novel source of pectin: Studies on antioxidant activity, functional group, and structural morphology." *Carbohydrate Polymer Technologies and Applications* 2: 100054. https://doi.org/10.1016/j.carpta.2021.100054

Lasprilla-Botero J., M. Álvarez-Láinez, and J. M. Lagaron. 2018. "The influence of electrospinning parameters and solvent selection on the morphology and diameter of polyimide nanofibers." *Materials Today Communications* 14: 1–9. https://doi.org/10.1016/j.mtcomm.2017.12.003

Lefebvre T., E. Destandau, and E. Lesellier. 2021. "Selective extraction of bioactive compounds from plants using recent extraction techniques: A review." *Journal of Chromatography A* 1635: 461770. https://doi.org/10.1016/j.chroma.2020.461770

Lerma-Torres J. M., A. Navarro-Ocaña, M. Calderón-Santoyo, L. Hernández-Vázquez, G. Ruiz-Montañez, and J. A. Ragazzo-Sánchez. 2019. "Preparative scale extraction of mangiferin and lupeol from mango (*Mangifera indica* L.) leaves and bark by different extraction methods." *Journal of Food Science and Technology* 56(10): 4625–31. https://doi.org/10.1007/s13197-019-03909-0

Li Y., D. Kong, Y. Fu, M. R. Sussman, and H. Wu. 2020. "The effect of developmental and environmental factors on secondary metabolites in medicinal plants." *Plant Physiology and Biochemistry* 148: 80–9. https://doi.org/10.1016/j.plaphy.2020.01.006

Lin L., Y. Zhu, and H. Cui. 2018. "Electrospun thyme essential oil/gelatin nanofibers for active packaging against *Campylobacter jejuni* in chicken". *LWT: Food Science and Technology* 97: 711–18. https://doi.org/10.1016/j.lwt.2018.08.015

Liu K., X. Zhang, L. Xie, M. Deng, H. Chen, J. Song, J. Long, X. Li, and J. Luo. 2021. "Lupeol and its derivatives as anticancer and anti-inflammatory agents: Molecular mechanisms and therapeutic efficacy." *Pharmacological Research* 164: 105373. https://doi.org/10.1016/j.phrs.2020.105373

Lu W., A. L. Kelly, and S. Miao. 2016. "Emulsion-based encapsulation and delivery systems for polyphenols." *Trends in Food Science and Technology* 47: 1–9. https://doi.org/10.1016/j.tifs.2015.10.015

Machado F. C., R. P. A. de Matos, F. L. Primo, A. C. Tedesco, P. Rahal, and M. F. Calmon. 2019. "Effect of curcumin-nanoemulsion associated with photodynamic therapy in breast adenocarcinoma cell line." *Bioorganic and Medicinal Chemistry* 27(9): 1882–90. https://doi.org/10.1016/j.bmc.2019.03.044

Matos I., A. F. Bento, R. Marcon, R. F. Claudino, and J. B. Calixto. 2013. "Preventive and therapeutic oral administration of the pentacyclic triterpene α, β-amyrin ameliorates dextran sulfate sodium-induced colitis in mice: The relevance of cannabinoid system." *Molecular Immunology* 54(3–4): 482–92. https://doi.org/10.1016/j.molimm.2013.01.018

Matsuura H. N., and A. G. Fett-Neto. 2015. "Plant alkaloids: Main features, toxicity, and mechanisms of action." In *Plant Toxins*, edited by P. Gopalakrishnakone, C. Carlini, and R. Ligabue-Braun, 1–15. Dordrecht: Springer. https://doi.org/10.1007/978-94-007-6728-7_2-1

Moraes-Lovison M., L. F. P. Marostegan, M. S. Peres, I. F. Menezes, M. Ghiraldi, R. A. F. Rodrigues, A. M. Fernandes, and S. C. Pinho. 2017. "Nanoemulsions encapsulating oregano essential oil: Production, stability, antibacterial activity and incorporation in chicken pâté." *LWT - Food Science and Technology* 77: 233–40. https://doi.org/10.1016/j.lwt.2016.11.061

Muhamad I. I., N. D. Hassan, S. N. H. Mamat, N. M. Nawi, W. A. Rashid, and N. A. Tan. 2017. "Extraction technologies and solvents of phytocompounds from plant materials: Physicochemical characterization and identification of ingredients and bioactive compounds from plant extract using various instrumentations." In *Ingredients Extraction by Physicochemical Methods in Food*, edited by A. M. Grumezescu and A. M. Holban, 523–60. Cambridge: Academic Press. https://doi.org/10.1016/b978-0-12-811521-3.00014-4

Patil S. P., and S. T. Kumbhar. 2018. "Evaluation of terpene-rich extract of *Lantana camara* L. leaves for antimicrobial activity against mycobacteria using Resazurin Microtiter Assay (REMA)". *Beni-Suef University Journal of Basic and Applied Sciences* 7(4): 511–15. https://doi.org/10.1016/j.bjbas.2018.06.002

Pogorzelska-Nowicka E., A. G. Atanasov, J. Horbańczuk, and A. Wierzbicka. 2018. "Bioactive compounds in functional meat products." *Molecules* 23(2): 1–19. https://doi.org/10.3390/molecules23020307

Prieto C., C. M. M. Duarte, and L. Calvo. 2017. "Performance comparison of different supercritical fluid extraction equipments for the production of vitamin E in polycaprolactone nanocapsules by supercritical fluid extraction of emulsions." *Journal of Supercritical Fluids* 122: 70–8. https://doi.org/10.1016/j.supflu.2016.11.015

Prieto C., and J. M. Lagaron. 2020. "Nanodroplets of docosahexaenoic acid-enriched algae oil encapsulated within microparticles of hydrocolloids by emulsion electrospraying assisted by pressurized gas." *Nanomaterials* 10(2): 270. https://doi.org/10.3390/nano10020270

Puppi D., and F. Chiellini. 2018. "Drug release kinetics of electrospun fibrous systems." In *Core-Shell Nanostructures for Drug Delivery and Theranostics*, edited by M. L. Focarete and A. Tampieri, 349–74. Sawston: Woodhead Publishing. https://doi.org/10.1016/B978-0-08-102198-9.00012-0

Radünz M., H. C. S. Hackbart, N. P. Bona, N. S. Pedra, J. F. Hoffmann, F. M. Stefanello, and E. D. R. Zavarese. 2020b. "Glucosinolates and phenolic compounds rich broccoli extract: Encapsulation by electrospraying and antitumor activity against glial tumor cells." *Colloids and Surfaces B: Biointerfaces* 192: 111020. https://doi.org/10.1016/j.colsurfb.2020.111020

Radünz M., H. C. S. Hackbart, T. M. Camargo, C. F. P. Nunes, F. A. P. de Barros, J. Dal Magro, P. J. S. Filho, E. A Gandra, A. L Radünz, and E. R. Zavarese. 2020a. "Antimicrobial potential of spray drying encapsulated thyme (*Thymus vulgaris*) essential oil on the conservation of hamburger-like meat products." *International Journal of Food Microbiology* 330: 108696. https://doi.org/10.1016/j.ijfoodmicro.2020.108696

Ramos-Hernández J. A., J. A. Ragazzo-Sanchez, M. Calderón-Santoyo, R. I. Ortiz-Basurto, C. Prieto, and J. M. Lagaron. 2018b. "Use of electrosprayed *Agave* fructans as nanoencapsulating hydrocolloids for bioactives." *Nanomaterials* 8(11): 1–11. https://doi.org/10.3390/nano8110868

Ramos-Hernández J. A., J. M. Lagarón, M. Calderón-Santoyo, C. Prieto, and J. A. Ragazzo-Sánchez. 2020. "Enhancing hygroscopic stability of *Agave* fructans capsules obtained by electrospraying." *Journal of Food Science and Technology* 58(4): 1593–603. https://doi.org/10.1007/s13197-020-04672-3

Ramos-Hernández J. A., M. Calderón-Santoyo, A. Burgos-Hernández, J. S. García-Romo, A. Navarro-Ocaña, M. G. Burboa-Zazueta, E. Sandoval-Petris, and J. A. Ragazzo-Sánchez. 2021. "Antimutagenic, antiproliferative and antioxidant properties of sea grape leaf extract fractions (*Coccoloba uvifera* L.)." *Anti-Cancer Agents in Medicinal Chemistry* 21(16): 2250–57. https://doi.org/10.2174/1871520621999210104201242

Ramos-Hernández J. A., M. Calderón-Santoyo, A. Navarro-Ocaña, J. C. Barros-Castillo, and J. A. Ragazzo-Sánchez. 2018a. "Use of emerging technologies in the extraction of lupeol, α-amyrin and β-amyrin from sea grape (*Coccoloba uvifera* L.)." *Journal of Food Science and Technology* 55(7): 2377–83. https://doi.org/10.1007/s13197-018-3152-8

Rathore P., and J. D. Schiffman. 2021. "Beyond the single-nozzle: Coaxial electrospinning enables innovative nanofiber chemistries, geometries, and applications." *ACS Applied Materials and Interfaces* 13(1): 48–66. https://doi.org/10.1021/acsami.0c17706

Rivera-Aguilar J. O., M. Calderón-Santoyo, E. M. González-Cruz, J. A. Ramos-Hernández, and J. A. Ragazzo-Sánchez. 2020. "Encapsulation by electrospraying of anticancer compounds from jackfruit extract (*Artocarpus heterophyllus* Lam): Identification, characterization and antiproliferative properties." *Anti-Cancer Agents in Medicinal Chemistry* 21(4): 523–31. https://doi.org/10.2174/1871520620666200804102952

Ruiz-Montañez G., A. Burgos-Hernández, M. Calderón-Santoyo, C. M. López-Saiz, C. A. Velázquez-Contreras, A. Navarro-Ocaña, and J. A. Ragazzo-Sánchez. 2015. "Screening antimutagenic and antiproliferative properties of extracts isolated from jackfruit pulp (*Artocarpus heterophyllus* Lam)." *Food Chemistry* 175: 409–16. https://doi.org/10.1016/j.foodchem.2014.11.122

Ruiz-Montañez G., M. Calderón-Santoyo, D. Chevalier-Lucia, L. Picart-Palmade, D. E. Jimenez-Sánchez, and J. A. Ragazzo-Sánchez. 2019. "Ultrasound-assisted microencapsulation of jackfruit extract in eco-friendly powder particles: Characterization and antiproliferative activity." *Journal of Dispersion Science and Technology* 40(10): 1507–15. https://doi.org/10.1080/01932691.2019.1566923

Santana-Gálvez J., L. Cisneros-Zevallos, and D. A. Jacobo-Velázquez. 2019. "A practical guide for sesigning effective nutraceutical combinations in the form of foods, beverages, and dietary supplements against chronic degenerative diseases." *Trends in Food Science and Technology* 88: 179–93. https://doi.org/10.1016/j.tifs.2019.03.026

Sarabandi K., P. Gharehbeglou, and S. M. Jafari. 2020. "Spray-drying encapsulation of protein hydrolysates and bioactive peptides: Opportunities and challenges." *Drying Technology* 38(5–6): 577–95. https://doi.org/10.1080/07373937.2019.1689399

Shah M. R., M. Imran, and S. Ullah. 2017. "Nanoemulsions." In *Lipid-Based Nanocarriers for Drug Delivery and Diagnosis*, 111–37. Norwich: William Andrew Publishing. https://doi.org/10.1016/B978-0-323-52729-3.00004-4

Sharma S., S. F. Cheng, B. Bhattacharya, and S. Chakkaravarthi. 2019. "Efficacy of free and encapsulated natural antioxidants in oxidative stability of edible oil: Special emphasis on nanoemulsion-based encapsulation." *Trends in Food Science and Technology* 91: 305–18. https://doi.org/10.1016/j.tifs.2019.07.030

Shin K., H. Choi, S. K. Song, J. W. Yu, J. Y. Lee, E. J. Choi, D. H. Lee, S. H. Do, and J. W. Kim. 2018. "Nanoemulsion vehicles as carriers for follicular delivery of luteolin." *ACS Biomaterials Science and Engineering* 4(5): 1723–9. https://doi.org/10.1021/acsbiomaterials.8b00220

Siacor F. D. C., K. J. A. Lim, A. A. Cabajar, C. F. Y. Lobarbio, D. J. Lacks, and E. B. Taboada. 2020. "Physicochemical properties of spray-dried mango phenolic compounds extracts." *Journal of Agriculture and Food Research* 2: 100048. https://doi.org/10.1016/j.jafr.2020.100048

Singh M., and Y. Shukla. 2015. "Combinatorial approaches utilizing nutraceuticals in cancer chemoprevention and therapy: A complementary shift with promising acuity." In *Genomics, Proteomics and Metabolomics in Nutraceuticals and Functional Foods*, edited by D. Bagchi, A. Swaroop, and M. Bagchi, 185–217. New Jersey: John Wiley & Sons. https://doi.org/10.1002/9781118930458.ch15

Singh Y., J. G. Meher, K. Raval, F. A. Khan, M. Chaurasia, N. K. Jain, and M. K. Chourasia. 2017. "Nanoemulsion: Concepts, development and applications in drug delivery." *Journal of Controlled Release* 252: 28–49. https://doi.org/10.1016/j.jconrel.2017.03.008

Soleimanifar M., S. M. Jafari, and E. Assadpour. 2020. "Encapsulation of olive leaf phenolics within electrosprayed whey protein nanoparticles; production and characterization." *Food Hydrocolloids* 101: 105572. https://doi.org/10.1016/j.foodhyd.2019.105572

Strati I. F., and V. Oreopoulou. 2014. "Recovery of carotenoids from tomato processing by-products - A review." *Food Research International* 65(C): 311–21. https://doi.org/10.1016/j.foodres.2014.09.032

Talegaonkar S., S. Pandey, N. Rai, P. Rawat, H. Sharma, and N. Kumari. 2016. "Exploring nanoencapsulation of aroma and flavors as new frontier in food technology". In *Encapsulations*, edited by A. M. Grumezescu, 47–88. Cambridge: Academic Press. https://doi.org/10.1016/b978-0-12-804307-3.00002-8

Tomaz I., L. Maslov, D. Stupić, D. Preiner, D. Ašperger, and J. K. Kontić. 2016. "Recovery of flavonoids from grape skins by enzyme-assisted extraction." *Separation Science and Technology* 51(2): 255–68. https://doi.org/10.1080/01496395.2015.1085881

Torkamani A. E., Z. A. Syahariza, M. H. Norziah, A. K. M. Wan, and P. Juliano. 2018. "Encapsulation of polyphenolic antioxidants obtained from *Momordica charantia* fruit within zein/gelatin shell core fibers via coaxial electrospinning." *Food Bioscience* 21: 60–71. https://doi.org/10.1016/j.fbio.2017.12.001

Vázquez-González Y., J. A. Ragazzo-Sánchez, and M. Calderón-Santoyo. 2020. "Characterization and antifungal activity of jackfruit (*Artocarpus heterophyllus* Lam) leaf extract obtained using conventional and emerging technologies." *Food Chemistry* 330: 127211. https://doi.org/10.1016/j.foodchem.2020.127211

Vergara C., M. T. Pino, O. Zamora, J. Parada, R. Pérez, M. Uribe, and J. Kalazich. 2020. "Microencapsulation of anthocyanin extracted from purple flesh cultivated potatoes by spray drying and its effects on *in vitro* gastrointestinal digestion." *Molecules* 25(3): 722. https://doi.org/10.3390/molecules25030722

Wang W. H., H. Y. Chuang, C. H. Chen, W. K. Chen, and J. J. Hwang. 2016. "Lupeol acetate ameliorates collagen-induced arthritis and osteoclastogenesis of mice through improvement of microenvironment." *Biomedicine and Pharmacotherapy* 79(155): 231–40. https://doi.org/10.1016/j.biopha.2016.02.010

Wangensteen H., D. Diallo, and B. S. Paulsen. 2015. "Medicinal plants from Mali: Chemistry and biology." *Journal of Ethnopharmacology* 176: 429–37. https://doi.org/10.1016/j.jep.2015.11.030

Weiss J., P. Takhistov, and D. J. McClements. 2006. "Functional materials in food nanotechnology." *Journal of Food Science* 71(9): 107–16. https://doi.org/10.1111/j.1750-3841.2006.00195.x

Wen P., M. H. Zong, R. J. Linhardt, K. Feng, and H. Wu. 2017. "Electrospinning: A novel nano-encapsulation approach for bioactive compounds." *Trends in Food Science and Technology* 70: 56–68. https://doi.org/10.1016/j.tifs.2017.10.009

Wink M. 2015. "Modes of action of herbal medicines and plant secondary metabolites." *Medicines* 2(3): 251–86. https://doi.org/10.3390/medicines2030251

Xu C. C., B. Wang, Y. Q. Pu, J. S. Tao, and T. Zhang. 2017. "Advances in extraction and analysis of phenolic compounds from plant materials." *Chinese Journal of Natural Medicines* 15(10): 721–31. https://doi.org/10.1016/S1875-5364(17)30103-6

Xu C. C., B. Wang, Y. Q. Pu, J. S. Tao, and T. Zhang. 2018. "Techniques for the analysis of pentacyclic triterpenoids in medicinal plants." *Journal of Separation Science* 41(1): 6–19. https://doi.org/10.1002/jssc.201700201

Zeković Z. P., S. D. Filip, S. S. Vidović, D. S. Adamović, and A. M. Elgndi. 2015. "Basil (*Ocimum basilicum* L.) essential oil and extracts obtained by supercritical fluid extraction." *Acta Periodica Technologica* 46: 259–69. https://doi.org/10.2298/APT1546259Z

Zhang R., T. Belwal, L. Li, X. Lin, Y. Xu, and Z. Luo. 2020. "Recent advances in polysaccharides stabilized emulsions for encapsulation and delivery of bioactive food ingredients: A review." *Carbohydrate Polymers* 242: 116388. https://doi.org/10.1016/j.carbpol.2020.116388

Ziaee A., A. B. Albadarin, L. Padrela, T. Femmer, E. O'Reilly, and G. Walker. 2019. "Spray drying of pharmaceuticals and biopharmaceuticals: Critical parameters and experimental process optimization approaches." *European Journal of Pharmaceutical Sciences* 127: 300–18. https://doi.org/10.1016/j.ejps.2018.10.026

CHAPTER 9

Bioaccessibility and Stability of Plant Secondary Metabolites in Pharmaceutical and Food Matrices

Javier Darío Hoyos-Leyva,
Iván Luzardo-Ocampo, and
Andrés Chávez-Salazar

CONTENTS

DOI: 10.1201/9781003166535-9

9.1 INTRODUCTION

Plant secondary metabolites are compounds of considerable interest due to their natural protective functions in the plant and their nutraceutical effect on the human body. There is a great diversity of plant secondary metabolites, grouped as phenolic compounds (simple phenols, stilbenes, flavonoids, polyphenols, furanocoumarins, isoflavonoids), terpenes (monoterpenes, sesquiterpenes, diterpenes, triterpenes, among others), and nitrogen- and sulfur-containing compounds (alkaloids, cyanogenic glucosides, phytochelatins, and glucosinolates) (Badyal et al. 2020). Chief among the plant secondary metabolites of health interest are the phenolic compounds, which are a diverse group of phytochemicals with antioxidant activity and a positive impact on the prevention of numerous chronic diseases, such as diabetes, obesity, cardiovascular diseases, and neurodegenerative disorders. Additionally, phenolic compounds play an important role in the sensory quality of food, both natural and processed products.

The absorption of the plant secondary metabolites in the human body depends on their bioaccessibility, which can be assessed by *in vitro* and *in vivo* methods, the most frequently used methods being simulations of the human digestive process (Alminger et al. 2014). The *in vitro* production of these metabolites has been studied due to some production advantages, such as reliability and predictability, while the bioaccessibility of the selected compound can be higher than its original plant matrix as potential interferences derived from the plant matrix are eliminated (Karuppusamy 2009). The plant secondary metabolites are easily isolated and purified from the plant cells and tissue cultures using controlled conditions. Isolation and purification of plant secondary metabolites cause the exposure of its chemical structures to oxidation-promoting factors. The factors that have been reported to increase the oxidation of secondary metabolites are light, temperature, pH, moisture, and reactions of the compounds of interest with other molecules present in a natural or processed product. Purified plant secondary metabolites can be used as ingredients or bioactive compounds in food and pharmaceutical products, where they can interact with other matrix compounds resulting in different bioaccessibility levels. Due to their instability and low bioaccessibility, plant secondary metabolites have been extracted and combined within several macro-, micro-, and nano-encapsulation methods to guarantee their bioactivity in targeted organs of the consumer.

Bioaccessibility has been defined as the amount of an ingested compound released from a food or pharmaceutical matrix in the gastrointestinal lumen which is available for intestinal absorption (Dima et al. 2020). Additionally, bioaccessibility includes the chemical changes by digestive enzymes into compounds ready for absorption into intestinal epithelium.

The bioaccessibility and stability of plant secondary metabolites depend on their physicochemical characteristics, affecting their chemical or physical interaction with other components in a food or pharmaceutical matrix. Additionally, the processing conditions can positively or negatively affect both bioaccessibility and stability due to unit operations involving temperature, pH changes, light exposure of the raw materials, and even chemical reactions. The extraction and processing techniques disrupting plant tissues and achieving denaturation of plant secondary metabolites complexes with other matrix components increases the bioaccessible fraction. However, thermal treatments can cause isomerization or other non-desirable chemical changes (Alminger et al. 2014).

Shahidi and Pan (2021) reviewed the influence of food matrix and food processing on the chemical interaction and bioaccessibility of dietary phytochemicals, indicating that the presence of components, such as proteins, carbohydrates, lipids, or even water from food or pharmaceutical matrices, can react with plant secondary metabolites, affecting their bioaccessibility.

The food and pharmaceutical matrices can be mainly liquid, solid, or an emulsion, presented as a foam, suspension, or a gel, among other physical structures, with multiple components that may interact by physical, chemical or both processing factors with these metabolites, causing variations in their bioaccessibility and stability (Parada and Aguilera 2007; Scholz and Williamson 2007; Yang et al. 2011). For instance, the presence of sugars in the food or pharmaceutical matrix can improve the bioaccessibility of metabolites, such as carotenoids (β-carotene), which exhibit a high bioaccessibility value in the presence of sucrose > glucose or fructose (Liu et al. 2020). Similarly, the flavan-3-ols from tea are absorbed faster in the gut after ingestion of a sugary drink, reducing the rate of gastric emptying (Borges et al. 2010). The bioaccessibility of bioactive cocoa flavan-3-ols can vary significantly depending on whether consumed in a matrix containing sugar or milk (Schramm et al. 2003; Neilson et al. 2010). Dietary fiber can affect the bioaccessibility of bioactive molecules through a series of mechanisms, including its reduced release from the cellular and tissue location inside the plant that contains them. The reduction of metabolite release is caused by their trapping by dietary fiber during their passage through the intestinal lumen or polysaccharide binding, requiring subsequent hydrolysis to be absorbed (Palafox-Carlos et al. 2011).

This chapter presents a review of the bioaccessibility and stability of plant secondary metabolites focused on the methods reported in the literature regarding food and pharmaceutical matrices.

9.2 DEFINITIONS AND METHODS TO TEST BIOACCESSIBILITY AND STABILITY OF PLANT SECONDARY METABOLITES

The definition of bioaccessibility has been extensively covered in recent years and involves the ability of particular metabolites to separate from the food or pharmaceutical matrix and be transported within the gastrointestinal fluids in a suitable form to be further absorbed (McClements 2020). Although the classical definition of bioaccessibility is as a property of metabolites generated during digestion, recently McClements and Peng (2020) suggested that bioaccessibility (B) is part of bioavailability (BA) in an equation as follows:

$$BA = B \times A \times D \times M \times E \tag{9.1}$$

where BA indicates bioavailability, A refers to absorption, D is the distribution of the metabolite, M is the metabolism of the metabolite, and E is its excretion. Bioavailability is usually defined as the metabolite fraction that is absorbed and available for its utilization in physiological functions and storage, although bioavailability does not necessarily indicate whether the substance is bioactive (Rousseau et al. 2020). Absorption considers the fraction of the bioaccessible form that is absorbed within the organism; distribution is the fraction of the metabolite at the site of action that has been distributed between several tissues in the human body; metabolism denotes the fraction that remains in an active form even after the physiological impact of enzymes or chemical components; and excretion of the substance depends on the way the substance is removed from the organism in the form of urine or feces. In this sense, bioaccessibility is a critical step for further evaluating bioavailability and the ADME parameters. Notably, most of these parameters critically depend on the initial substance concentration, the residence time, and chemical–biological interactions between the metabolites and physiological factors, such as temperature, pH, and enzyme activity, among other parameters (Luzardo-Ocampo et al. 2017).

Some of the most studied plant secondary metabolites, namely phenolic compounds, display chemical properties which greatly influence their bioaccessibility. For instance, several flavonoids are originally glycosylated, forming polymeric forms that require their proper deglycosylation to be absorbed. At the same time, other kinds of phenolics (e.g., conjugated flavonoids) cannot be degraded by human gastrointestinal enzymes and potentially undergo microbial fermentation, producing important metabolites with a wide range of biological properties (Ribas-Agustí et al. 2018). It has been reported that most plant secondary metabolites could be retained in the non-digestible fraction, reaching the colon to produce fermentation-derived metabolites such as short-chain fatty acids (SCFAs) or SCFA-derivatives, among other components, depending on the original composition of the food or pharmaceutical matrix (Luzardo-Ocampo et al. 2020b) (Luzardo-Ocampo et al., 2020a).

Bioaccessibility supports the health-associated functionality of plant secondary metabolites as it partially ensures their ability to potentially cross the gastric or intestinal barrier to further exert their biologically active properties (Caicedo-Lopez et al. 2019). The ability of substances to cross the intestinal barrier can be quantified under several direct or indirect *in vitro* and *in vivo* approaches, defining their intestinal permeability.

Gastrointestinal models have been used to assess bioaccessibility and can be static or dynamic. The static methods are quick and low-cost methods, but the principal disadvantage is that they do not consider transport steps in the gastrointestinal tract. The dynamic methods require a high-tech system that should simulate all the digestion steps, which implies accurately simulating the effects of the mastication process, digestive fluids, and pH, among other conditions (Angelino et al. 2017). Alminger et al. (2014) reviewed the *in vitro* models for studying the digestion and bioaccessibility of plant secondary metabolites and reported that these models should simulate the oral phase, gastric phase, and small and large intestine steps, as these metabolites could be affected by physiological conditions, such as the presence of digestive enzymes and low pH values (gastric phase, pH 2–3.5). Other reviews have been focused on *in vitro* models to test the bioaccessibility of plant secondary metabolites (Etcheverry et al. 2012; Rein et al. 2013; Angelino et al. 2017; Furtado et al. 2017; Dima et al. 2020; Ištuk et al. 2020; Câmara et al. 2021).

Bioaccessibility is an exclusively *in vitro* parameter, whereas bioavailability requires *in vivo* approaches. There have been some attempts to calculate potential bioavailability (Ariza et al. 2018; Celep et al. 2018), but, based on the nature of the proposed assays, bioavailability was not measured, rather than the relationship between those compounds contained in selected intestinal fractions was determined. Proposals for calculating potential bioavailability at several steps of the *in vitro* gastrointestinal digestion are shown in Table 9.1

Bioaccessibility quantifications need proper intestinal digestion simulations or experimental procedures, but some authors have done quantification by using centrifugation or mechanical separation of compounds (Tian et al. 2021), which is highly questionable as no permeation barrier was used. Rather than the amount of compounds at the basolateral or apical sides of the small intestine, bioavailability refers to the ability of those compounds to reach the systemic circulation with a potential to exert a biological effect, considering the ADME pharmacokinetic stages (Shahidi and Peng 2018).

There are several proposals for calculating bioaccessibility. Some authors have suggested that bioaccessibility (B) is the ratio between the amount of particular components at a specific gastrointestinal stage and the number of compounds from the original undigested matrix (D'Antuono et al. 2015):

$$\mathrm{B}\ (\%) = (\mathrm{C_f}/\mathrm{C_i}) \times 100 \tag{9.2}$$

TABLE 9.1 Bioaccessibility and Potential Bioavailability Calculations During *in vitro* Gastrointestinal Digestion

Equation	Variables
$B_G = \dfrac{GF_{out} + GF_{in}}{CI} \cdot 100\%$	**BG**: Gastric bioaccessibility; $\mathbf{GF_{out}}$: amount of compound outside of the gastric membrane; $\mathbf{GF_{in}}$**:** amount of compound inside the gastric membrane; **CI:** Initial concentration of the compound in the original extract.
$B_i = \dfrac{IF_{out} + IF_{in}}{CI} \cdot 100\%$	**BI**: Intestinal bioaccessibility; $\mathbf{IF_{out}}$**:** amount of compound outside the intestinal membrane; $\mathbf{IF_{in}}$: amount of compound inside the intestinal membrane; **CI**: Initial concentration of the compound in the original extract.
$BP_G = \dfrac{GF_{out}}{CI} \cdot 100\%$	$\mathbf{BP_G}$: gastric potential bioavailability.
$BP_i = \dfrac{IF_{in}}{CI} \cdot 100\%$	$\mathbf{BP_i}$: Intestinal potential bioavailability.

Adapted from Ariza et al. (2018).

where C_f refers to the amount of compound at the digestion stage (oral, gastric, or intestinal), and C_i is the initial amount from the original matrix. Although simple, this formula can be controversial since the number of released compounds (C_f) could exceed the quantified compounds from the undigested sample (C_i). Hence, an alternative formula has been proposed as follows (Luzardo-Ocampo et al. 2017):

$$B\ (\%) = (C_f - C_i/C_i) \times 100 \tag{9.3}$$

However, particularly for encapsulation systems, McClements and Peng (2020) suggested re-formulating the calculation of bioaccessibility, expressing the amount of digested compound relative to the encapsulation efficiency as not all the bioactive compounds from a matrix are fully digested. In addition, using the original formula may lead to bioaccessibilities greater than 100%, depending on the type of compounds being quantified and the methods used for their quantification. For instance, total polyphenols are usually quantified using the Folin-Ciocalteu method, but not all the extractable polyphenols can be detected under this procedure, and the concentration of the released compounds at the gastrointestinal system can easily exceed 100% (Caicedo-Lopez et al. 2019). However, since the Folin-Ciocalteu procedure is a spectrophotometric method highly susceptible to being affected by optic interferences, quantification of these compounds should use chromatographic systems, such as high-performance liquid chromatography (HPLC), that might be coupled to diode array detection (DAD) or mass spectrometry (MS) (Sánchez-Rangel et al. 2013).

Bioaccessibility is closely associated with the rate at which compounds cross the epithelial barrier, known as intestinal permeability (Artursson et al. 2001); this latter parameter can be measured indirectly through several *in vitro* or *in vivo* methods. Independently of the method, the most common way to quantify intestinal permeability was proposed by Lassoued et al. (2011) for drugs, calculating the apparent permeability coefficient (P_{app}, cm/s):

$$P_{app} = (dQ/dt)[1/(AC_0)] \tag{9.4}$$

where dQ/dt (mg/s) is the rate of transport of the drug or metabolite across the membrane, A (cm^2) is the surface area available for permeation, and C_0 (mg/mL) is the initial concentration of the drug outside the sac. As this equation is intended for dialysis membranes, *ex-vivo* adaptations can be conducted using everted rat sacs from several strains such as Wistar (Herrera-Cazares et al. 2021) or Sprague-Dawley (Vega-Rojas et al. 2021) rats, coupled to *in vitro* gastrointestinal methodologies (Campos-Vega et al. 2015). Recently, Mateer et al. (2016) used an *ex-vivo* method to assess the intestinal permeability of several drug models (dextran sodium sulfate or DSS, fluorescein isothiocyanate or FITC, and dexamethasone, among others), using sectioned everted small intestine from C57BL/6 mice into the jejunum, ileum, and colon, showing regional differences in the P_{app} values. For instance, a commonly used chemical substance to induce intestinal inflammation, dextran sodium sulfate (DSS), displayed values up to 2.5×10^{-5} cm/s at the jejunum, ~1.8-2.4×10^{-5} cm/s at the ileum, and up to ~1.3×10^{-5} cm/s at the colon. The same authors also evaluated the P_{app} values as indicated by Lassoued et al. (2011) but suggested the calculation of the cumulative concentration Q_t to be used in the dQ/dt factor as indicated:

$$Q_t = (C_t \times V_r) + (Q_{t\,sum} \times V_s) \tag{9.5}$$

where C_t is the concentration of the compound at the time *t*, V_r is the volume at the receiver side (basolateral side of the everted gut sac), $Q_{t\,sum}$ is the sum of all previous Q_t values at a specific time, and V_s is the volume sampled. Like P_{app}, the trans-electrical epithelial resistance (TEER) from *ex-vivo* tissues or cell cultures is also an assessed parameter to indirectly show the impact of drug or food matrices on the intestinal permeability (Nunes et al. 2016). The polarized human Caco-2 colon adenocarcinoma cell line has been widely used to evaluate TEER, considering that growing these cells for up to 21–24 days can change their phenotype to epithelial cells, serving as a model barrier in proper cell culture systems, such as Transwell® inserts (Luzardo-Ocampo et al. 2020). For this, a simple equation calculation, transforming the voltage values to $V \cdot cm^2$, gives information about the extent to which compounds can disrupt the barrier, resulting in a decreasing permeability.

The stability of a product containing plant secondary metabolites should be tested for both physical and chemical stability. The physical stability of the compounds can be estimated from integral entropy change during the shelf-life studies of a pharmaceutical or food matrix (Hoyos-Leyva et al. 2018b). The integral entropy can be calculated from the sorption isotherms of the product containing the plant secondary metabolites, implying an experimental run at different temperatures, generally under 65 °C (Hoyos-Leyva et al. 2018). The physical stability of ascorbic acid and almond oil encapsulated into taro starch spherical aggregates was assessed by thermodynamic analysis (Hoyos-Leyva et al. 2018a). The physical stability of microencapsulated compounds was related to their interaction with the wall material. The thermodynamic analysis allows the determination of the storage conditions such as temperature, relative humidity, and water activity in the product, resulting in prolonged stability of the plant secondary metabolites (Hoyos-Leyva et al. 2018b). Most stability studies of plant secondary metabolites are associated with shelf-life tests, implying the determination of the bioactivity and the physical changes during storage. Accelerated storage has been a widely used method to assess the shelf-life of foods and pharmaceuticals (Corradini and Peleg 2007), which consists of storing the product under extreme conditions of temperature, moisture, light, and gas concentrations.

The study of the stability of plant secondary metabolites, using antioxidant capacity methods as secondary indicators, commonly implies the measure from total phenolic content, the 2,2-diphenyl-1-picrylhydrazyl (DPPH) scavenging activity, 2,2'-azino-bis (3-ethylbenzothiazoline-6-sulfonic acid) (ABTS) scavenging activity, and the measure of the compound interest concentration before and after the product storage under desirable conditions. The stability is directly related to the bioactivity of the compound because it involves the desirable specific effect after exposure of the compound (Karaś et al. 2017).

9.3 BIOACCESSIBILITY AND STABILITY OF SECONDARY METABOLITES IN PHARMACEUTICAL MATRICES

Bioaccessibility of plant secondary metabolites depends on the pharmaceutical matrix and the mechanism of its degradation. These factors are related to the mechanism of the metabolite release. The release can be by mastication pressure on the capsule or the digestive conditions, inducing the release of bioactives from the matrix. Generally, the pharmaceutical matrices are composed of the bioactive plant secondary metabolite, a releasing controlling agent, the matrix former, matrix modifier, solubilizers and pH modifiers, lubricant and flow aids, supplementary coating to extend the lag time, and density modifiers. Due to the mixture of the compounds in the pharmaceutical matrix, bioaccessibility is critical to guarantee the biological properties of matrices, particularly pharmaceutical-like components, in order to deliver a health benefit. Table 9.2 shows

TABLE 9.2 Examples of Bioaccessibility Methods (BM) Used in Pharmaceutical Matrices and Plant Secondary Metabolites of Interest.

Product	**Process**	**Plant Secondary Metabolites**	**BM**	**Bioaccessibility**	**Reference**
Solid matrix					
Dried leaves extract	Decoction	Saponins, phenolic acids	Oral, gastric, and intestinal digestion	TPC: 97.61%; Tannins: 88.14 %; Saponins: 70.68 %	Jagadeesan et al. 2022
Encapsulated pine bark extract	Spray drying	Flavonoids, phenolic acids	Oral, gastric, and intestinal digestion	Oral: 87.1 $\pm$ 5 % Gastric: 58.5 $\pm$ 3 % Intestinal: 54.3 $\pm$ 1 %	Ferreira-Santos et al. 2021
Composites	Antisolvent precipitation	Resveratrol	Gastrointestinal digestion	17.5 % (resveratrol)–52.5 % (resveratrol composites)	Ji et al. 2020
Gold nanoparticles	Nanoprecipitation	Genistein	Dialysis	< 58 %	Zhang et al. 2015
		Hesperetin		< 60 % at 10 h, 99% until 80 h	Krishnan et al. 2017
Solid/gel					
Microparticles and emulsions	Electrostatic complexation Emulsification	β-carotene	Oral, gastric, and intestinal digestion	Microparticles >80 % Emulsions >70 %	Liu et al. 2018
Gel matrix					
Encapsulated β-carotene	Emulsification	β-carotene	Oral, gastric, and intestinal digestion	> 60 %	Qi, Zhang, and Wu 2020
Liquid matrix					
Fermented *Psidium guajava* leaf extract	Fermentation + extraction	Phenolic compounds	Gastric and intestinal phase	Total flavonoids: 50 % (gastric stage); <25 % (intestinal stage). TPC: >40 % (gastric stage), >50 % (intestinal stage)	Huang et al. 2021

TPC: Total phenolic compounds.

some examples of the matrix type (solid, liquid, solid/gel) and the plant secondary metabolites of pharmaceutical interest, and the method used to measure its bioaccessibility. Huang et al. (2021) reported the bioaccessibility of plant secondary metabolites from fermented *Psidium guajava* leaves. The leaves were fermented using a *Bacillus* sp. BS2 culture medium mixed with *Monascus anka* seed culture and the product obtained was dried, sieved, and extracted with 70% ethanol under ultrasonic treatment. Total phenolic compounds were more bioaccessible than total flavonoids (51.2% vs. 16.6%, respectively), and the low hydrophobicity of the total flavonoids was linked to their low bioaccessibility. Ellagic acid was found to be the most bioaccessible phenolic, followed by isoquercitrin and quercetin-3-*O*-β-D-xylopyranoside. An acute toxicity study showed no toxic effects on male ICR mice.

Most applications of the bioaccessibility of plant secondary metabolites in pharmaceutical matrices focus on encapsulation methods to bring these compounds to targeted gastrointestinal sites, producing a beneficial health effect. For instance, Ferreira-Santos et al. (2021) reported the bioaccessibility and bioactivity of an encapsulated pine bark polyphenolic extract for antimicrobial effect and prevention of oxidative stress against *Listeria innocua*, a harmless microorganism used as a surrogate for *Listeria monocytogenes*. The encapsulation process was conducted through spray drying, using maltodextrin as the wall material, whereas the extract was prepared with a water/methanol solvent (30:70 v/v). Although uniform capsules could not be obtained, the bioaccessibility displayed by total phenolic compounds was higher ($p<0.05$) compared with the non-encapsulated material (quantified mean ± standard error differences between the encapsulated material and the non-encapsulated material at different stages were oral: 15.04 ± 0.53%; gastric: 35.39 ± 0.68%; and intestinal: 34.78 ± 7.55% stages). The high bioaccessibility obtained as a result of the encapsulation process reflected high encapsulation efficiencies of selected components and derived properties such as total phenolic compounds (TPC): 67%; antioxidant capacity measure by ABTS: 98%; ferric-reducing antioxidant capacity (FRAP): 99%. In addition, TPC reached up to 200 mg GAE per g dry weight at the oral stage. Antisolvent precipitation is a method to produce composites, which improve the low bioaccessibility and water solubility of some plant secondary metabolites. Resveratrol-dietary fiber (DF) composites produce a better resveratrol bioaccessibility (300% higher) than the resveratrol used on its own (Ji et al. 2020).

Despite no membrane being used for intestinal digestion, Qi et al. (2020) reported the bioaccessibility of encapsulated β-carotene in oleogel-in-water Pickering emulsions (OPE), aiming to potentially increase the bioaccessibility of this carotenoid due to its health benefits. For this, cellulose nanocrystals were prepared as emulsifiers and mixed with 1 mg/mL β-carotene previously dissolved in soybean oil as a carrier. Two OPE systems (OPE-1 and OPE-2) used for encapsulation purposes were further evaluated at the digestion simulation. Bioaccessibility was calculated from the amount of β-carotene from the micelles at each stage of the gastrointestinal digestion vs. the initial amount from the undigested matrix. Results showed high β-carotene chemical stability, reaching > 80% of retention during up to five days of storage (and 60–70 % stability after 15 days). The two encapsulation processes displayed about 60 and 80% bioaccessibilities for OPE-1 and OPE-2, respectively.

A different β-carotene encapsulation procedure was assayed by Liu et al. (2018) to enhance its bioaccessibility. For this, both β-carotene-loaded emulsions (sodium caseinate) and microparticles (2% sodium caseinate and 2% alginate) were subjected to a simulated gastrointestinal digestion model without using membranes for the intestinal step. With the traditional calculation of bioaccessibility using the initial concentration of β-carotene in the samples (referred to as effective bioaccessibility), bioaccessibility was calculated using the remaining β-carotene in the digestive steps. As a result, higher β-carotene retention (80% vs. 70%) and bioaccessibility (>80%) were obtained from microparticles, compared with the emulsions.

The dialysis method was used to measure the release of the genistein (an isoflavonoid with potential antitumor efficacy) from poly (ε-caprolactone) (PCL) and a copolymer was synthesized from ε-caprolactone initiated by D-α-tocopheryl polyethylene glycol 1000 succinate (TPGS-β-PCL) nanoparticles (Zhang et al. 2015), which are involved in the measure of the *in vitro* release of the compound from the matrix. To produce that, the sample was resuspended in a releasing buffer (phosphate-buffered saline containing Tween 80). The release of the genistein was higher from TPGS-β-PCL (58%) than from PCL (48.9%) for 15 days. These authors concluded that the copolymer TPGS-β-PCL had a greater water permeability because of its hydrophilic characteristics.

Gold nanoparticles have been developed to improve the solubility, bioaccessibility, and bioavailability of hesperetin (5, 7, 3'-trihydroxy-4' methoxy flavanone) (Krishnan et al. 2017), a flavonoid of citrus fruits which occurs in its glycoside form hesperidin in nature. Hesperitin is recognized in Chinese traditional medicine for its antitumor effect. The *in vitro* release test showed that hesperitin can be up to 80 % bioaccessible in 10 h from the gold nanoparticles and up to 99 % at 60 h, demonstrating that the drug-loaded nanoparticles are a suitable carrier with a sustained release behavior.

9.4 BIOACCESSIBILITY AND STABILITY OF SECONDARY METABOLITES IN FOOD MATRICES

After consuming food containing plant secondary metabolites, chemical modifications occur due to the action of digestive enzymes, affecting the physicochemical properties of such metabolites in the digestive tract. There are soluble phenolic compounds in free and conjugated form, whose bioaccessibility will depend on their stability during the digestion process (Putnik et al. 2019). These compounds are well known for their poor bioavailability. For instance, only 5–10 % of these phenolics are absorbed, and the remaining 90–95%, pass toward the colon to be subjected to microbial metabolism and degradation by intestinal microbiota (Zhang et al. 2020). The bioaccessibility of the phenolic compounds can be improved by chemical and physical processes during food production. Ribeiro et al. (2019) found that the biosorption of plant secondary metabolites in *Saccharomyces cerevisiae* increases (up to 200% improvement of bioaccessibility) due to chemical modifications carried out by the microorganism.

Extensive research has been carried out testing the bioaccessibility and the bioavailability of plant secondary metabolites in food matrices using *in vitro* and *in vivo* models (Table 9.3). These methods facilitate the study of the impact of the digestion process on the chemical structure of food components. For example, Jagadeesan et al. (2022) reported the bioaccessibility of polyphenols from *Allmania nodiflora* leaves. Several extracts were obtained after conducting an *in vitro* gastrointestinal digestion without using an intestinal membrane. Different metabolites were identified, such as saponins, amino acids, hydroxyl acids, flavonoids, and phenolic compounds. TPC exhibited a 97.6% bioaccessibility, followed by tannins (88.1%), and saponins (70.7%). Also, a considerable inhibition of α-amylase and α-glucosidase, > 50% activity, was noted. Hence, the presence of rutin, caffeic acid, chlorogenic acid, saikosaponin, and catechin in *A. nodiflora* leaves might play a key role in managing hyperglycemia; inhibition of starch digestion enzymes is related to controlled postprandial hyperglycemia. Research conducted on the baobab (*Adansonia digitata*) fruit, rich

TABLE 9.3 Bioaccessibility Methods (BM) Used to Test Food Matrices Containing Plant Secondary Metabolites

Product	Plant Source	Process	Plant Secondary Metabolites	BM	Bioaccessibility	Reference
Liquid matrix						
Extract	Cactus berry (*Myrtillocactus geometrizans*)	Juice extraction	Betalain, phenolic compounds	Gastrointestinal digestion	Bx-glutamine 2 % Bx-dopamin 6 % Epigallocatechin-glucoside and kaempferol-7-O-neohesperidosede 50 %	Montiel-Sánchez et al. 2021
Extract	Yerba mate	Biosorption on *Saccharomyces cerevisiae*	Phenolic compounds	Gastrointestinal digestion	100 %	Ribeiro et al. 2019
Juice	Red dragon fruit	Juice extraction	Betanin	Gastrointestinal digestion	Gastric: 77.68 % Intestinal: 43.76 %	Choo et al. 2019
Fermented drink	Red dragon fruit	Juice extraction +fermentation	Betanin	Gastrointestinal digestion	Gastric: 86.44 Intestinal: 46.42 %	
Beverage	Green roasted coffeeYerba mate		Caffeoylglucose3-Caffeoylquinic acid Dicaffeoylquinic acid Caffeoyl-feruloylquinic acidCaffeoylshikimic acid Caffeoylquimic lactone rutin Quercitin glycosede Kaempferol rhamnoglucoside	Gastric digestion Gastrointestinal digestion	Gastric: 37-100 % Gastric+intestinal: 43-100 %	Baeza et al. 2018

(*Continued*)

TABLE 9.3 (CONTINUED) Bioaccessibility Methods (BM) Used to Test Food Matrices Containing Plant Secondary Metabolites

Product	Plant Source	Process	Plant Secondary Metabolites	BM	Bioaccessibility	Reference
Solid matrix						
Powder	*Mangifera indica* cv. Ataulfo	Lyophilized mango bagasse	Phenolic acidsGallic acid Benzophenones Xanthones (mangiferin, homomangiferin, and isomangiferin)	Gastrointestinal digestion	100% mangiferin 100% methyl gallate < 10% methyl gallate ester < 20% Octa-*O*-galloyl hexoside	Herrera-Cazares et al. 2021
Flour	Maize	Cooking	Campesterol Stigmasterol β-Sitosterol	Gastrointestinal digestion	Campesterol ~50% Stigmasterol ~50% β-Sitosterol < 50%	Hossain and Jayadeep 2020
	Red sorghum	Cooking	PhenolicsFlavonoids Condensed tannins	Gastrointestinal digestion	100%	Luzardo-Ocampo et al. 2020
Flour	Red sorghum	Nixtamalization	PhenolicsFlavonoids Condensed tannins	Gastrointestinal digestion	10 to 100%	
	White sorghum	Cooking	PhenolicsFlavonoids Condensed tannins	Gastrointestinal digestion	TPC and CT ~ 100 % TF > 50%	
	White sorghum	Nixtamalization	PhenolicsFlavonoids Condensed tannins	Gastrointestinal digestion	TPC > 50% CT 100% TF > 50%	
Biscuit	Horseradish (*Armoracia rusticana*)	Baking	(+)-CatechinSinapic acid 2-Hydroxycinnamic acidRutin	Gastrointestinal digestion	Total phenolics increase from 18 to 30%	Tomsone et al. 2020
Wheat bread	Green coffee	Baking	Phenolics Caffeine	Gastric digestion	Phenolics: 25% Caffeine: 30.7%	Świeca et al. 2017

TPC: total phenolic compounds; TF: total flavonoids; CT: condensed tannins.

in polyphenols with high antioxidant capacity, by *in vitro* digestion and extraction from both pulp and skin, found that compounds such as flavonoids, particularly quercetin, proanthocyanidin, proanthocyanidins B1 and B2, were highly bioaccessible (Ismail et al. 2021). Similarly, Tomas et al. (2021) analyzed the metabolome composition before and after *in vitro* gastrointestinal digestion of four brassicaceous vegetables (kale, red cabbage, kohlrabi, and radish). A significant diversity (470 compounds) of secondary metabolites was found. It was reported that, among polyphenols, flavonoids were the most representative ones (180 compounds, which included anthocyanins, flavones, flavonols, and other flavonoids), followed by phenolic acids (68 compounds, mainly hydroxycinnamic and hydroxybenzoic acids), compounds whose structure was based on phenolic acid or non-flavonoids (i.e., alkyl-phenols and alkyl methoxy-phenols and derivatives of tyrosol), and lignans. Twenty-two glucosinolates were registered, including gluconapine, glucoraphanin, glucobrassicin, and 4-hydroxyglucobrassicin. Lignans exhibited the highest bioaccessibility (14%), followed by tyrosol and flavonoid derivatives (on average, 9 and 8%, respectively). Differences might be attributable to the plant species, where comparatively lower bioaccessibility values were found in red cabbage regardless of the chemical class of the compound.

Other plant species, such as cactus berry (*Myrtillocactus geometrizans*), have been studied. Montiel-Sánchez et al. (2021) evaluated the cactus berry fruits and identified 43 metabolites by High-performance liquid chromatography coupled to diode array detection, electrospray ionization and quadrupole time-of-flight (HPLC-DAD-ESI-QTOF) (including eight betaxanthins, eight betacyanins, 13 flavonoids, and six phenolic acids). Phylocactin and isorhamnetin rhamnosyl-rutinoside were the most abundant metabolites (5876 and 396 μg/g dry weight), showing a bioaccessibility of 16 and 21%, respectively. Likewise, the fruit pulp exhibited greater antioxidant activity than the other fruit parts because of the absorbance capacity of oxygen radicals (27 mM Trolox equivalents). On the other hand, the antihyperglycemic activity was higher in the skin and pulp tissues (85% inhibition of α-glucosidase and 8% of α-amylase activities).

Many processing techniques have been investigated to protect plant secondary metabolites during gastrointestinal digestion. Carrasco-Sandoval et al. (2021) studied nanoparticles based on zein (ZN) and zein/polysaccharides (ZNP) to co-encapsulate different types of phenolic compounds and evaluate the effect on their bioaccessibility. Two phenolic acids, two glycosylated flavonoids, and three aglycone flavonoids were co-encapsulated.

The encapsulation efficiency of phenolic compounds loaded in ZN was in the range from 95.3 to 98.5 %. On the other hand, the ZNP-based samples significantly increased the encapsulation efficiency of phenolic acids and glycosylated flavonoids. The bioaccessibility of phenolic compounds loaded in ZN increased by at least 50.2%. The improved bioaccessibility of glycosylated phenolic acids and flavonoids was observed when coated with alginate, although the highest bioaccessibility of polyphenols was observed when loaded in ZNP (> 65.8%).

Another encapsulation technique commonly used in the food industry is the spray-drying process. Szabo et al. (2021) carried out qualitative and quantitative analysis by high-performance liquid chromatography (HPLC) of carotenoid compounds from tomatoes and tomato compounds microencapsulated by spray drying into an oil-in-water emulsion to encapsulate carotenoids with linseed oil as a vehicle and a binary mixture of gum arabic and maltodextrin (1:1 w/w) as the wall materials. The microcapsules were subjected to an *in vitro* simulated gastrointestinal digestion process to assess the bioaccessibility of the carotenoids. The results showed considerable degradation of lycopene during the spray-drying process and constant degradation of β-carotene throughout the gastric phase of simulated digestion. Bioaccessibility depends on the digestive step that has been simulated. Total phenolics acids, flavonoids, and polyphenols are less bioaccessible in the intestinal phase than in the gastric phase (Ovando-Martínez et al. 2018).

Once the metabolites have been obtained by different techniques, it is necessary to evaluate the stability over time and the effect that different storage and processing conditions may exert on them. An example of that was reported by Cassani et al. (2020), who explored algae as promising matrices to generate added value by incorporating them into food, as a functional alternative as a result of their secondary metabolite composition. The authors suggest that the pre-processing operations, extraction processes, and long-term storage play an important role in any increase or decrease of the phlorotannin content. This research highlighted that phlorotannins are exposed to the human gastrointestinal tract, where the physiological pH and digestive enzymes can significantly affect their stability and biological activity. In addition, compounds such as tannins have been associated with astringency and bitter taste, so effective doses of phlorotannins can adversely affect the sensory attributes of food products, limiting their use. Due to this series of drawbacks, it is necessary to apply intelligent strategies to develop products that provide the necessary protective mechanisms to maintain the active molecular form of the metabolites of interest from product fabrication to consumption, which control their release upon arrival at the intestine. Temperature is a factor of critical importance during the processing of foods, as high temperatures can induce metabolite degradation. Zorenc et al. (2017) studied the impact of different temperatures (4 °C, room temperature, 50, 75, and 100 °C) on the stability of blueberry purées. TPC decreased between 11 and 32% at the highest temperature. On the other hand, Perestrelo et al. (2017) analyzed the effect of temperature, storage period, and glucose content on the formation of furan derivatives (FD) in fortified Madeira wines by means of headspace solid-phase microextraction (HS-SPME) combined with gas chromatography and quadrupole mass spectrometry (GC-qMS), and tandem packed sorbent microextraction (MEPS) with ultra-high pressure liquid chromatography coupled to photodiode array detection (UPLC-PDA). MEPS/UPLC-PDA showed a more effective extraction of FD compared with HS-SPME/GC-qMS. The MEPS/UPLC-PDA extraction caused an increase in the concentration of FD during storage, regardless of the storage temperature. The increase in FD could be favored by increases in temperature, being more significant at 55 °C. Chen et al. (2020) compared the degradation kinetics of red raspberry anthocyanins in both fresh fruit and juice extract during storage at 37 °C. The two main anthocyanins, cyanidin-3-glucoside (Cy-3-glc) and cyanidin-3-soporoside (Cy-3-soph), degraded faster in the juice than in the fresh fruit, and the authors attributed this effect to the components co-existing in the juice. However, these intrinsic factors did not alter the main degradation pathways of anthocyanins. It was hypothesized that anthocyanins in the juice system follow similar degradation pathways in fresh fruit. The degradation of anthocyanins into small colorless molecular components led to loss of original color in red raspberry juice and loss of antioxidant activity. The antioxidant activity of the low-molecular-weight compounds was lower than the antioxidant activity of the fresh fruit. Kuck et al. (2017) assessed the stability of the microparticles of an aqueous extract isolated from the skin of Bordo grapefruit produced by atomization and lyophilization with polydextrose (5%) and partially hydrolyzed guar gum (5%), analyzed under conditions of accelerated shelf-life (75 and 90% relative humidity, at 35, 45 and 55 °C for 35 days) and simulated gastrointestinal digestion. The temperature had a significant effect on the decrease in phenol content, with retentions that varied from 82.5 % to 93.5 %. The retention of total monomeric anthocyanins ranged from 3.9 % to 42.3 %. The antioxidant capacity exhibited a final retention of 38.5 % to 59.5 %. In the simulated gastrointestinal digestion, the authors observed a maximum release of phenolic compounds in the intestinal phase (90.6 and 94.9 % from spray-dried and lyophilized powders, respectively), which agreed with the highest antioxidant capacity values (69.4 % and 67.8 % from spray-dried and lyophilized powders, respectively).

9.5 CONCLUSION

Bioaccessibility and stability of plant secondary metabolites are critical parameters to be considered in the bioproducts design, including food and pharmaceutical matrices; the major metabolites of interest in the food products are the phenolic compounds due to their antioxidant activity and their functional properties, such as color, taste, and fragrance modification. Studies about bioaccessibility and stability focus on the *in vitro* simulation of gastrointestinal digestion, including the oral phase. The pharmaceutical bioproducts are tested by *in vitro* release methods. Quantification of the bioaccessibility of secondary metabolites relates to the concentration of the compound at any digestive stage relative to the initial concentration, which implies improvement of the techniques to isolate the compounds from complex matrices and their quantification before the digestion analysis. Research into the stability of plant secondary metabolites is directly related to bioaccessibility and bioavailability due to the importance of the preservation of the compounds' bioactivity. The scientific challenge for the future, among other factors, is related to the improvement of the latter three key characteristics.

REFERENCES

Alminger M., A. M. Aura, T. Bohn, C. Dufour, S. N. El, A. Gomes, S. Karakaya, M. C. Martínez-Cuesta, G. J. McDougall, T. Requena, and C. N. Santos. 2014. "*In vitro* models for studying secondary plant metabolite digestion and bioaccessibility." *Comprehensive Reviews in Food Science and Food Safety* 13(4): 413–36. https://doi.org/10.1111/1541-4337.12081

Angelino D., M. Cossu, A. Marti, M. Zanoletti, L. Chiavaroli, F. Brighenti, D. Del Rio, and D. Martini. 2017. "Bioaccessibility and bioavailability of phenolic compounds in bread: A review." *Food and Function* 8(7): 2368–93. https://doi.org/10.1039/c7fo00574a

Ariza M. T., P. Reboredo-Rodríguez, L. Cervantes, C. Soria, E. Martínez-Ferri, C. González-Barreiro, B. Cancho-Grande, M. Battino, and J. Simal-Gándara. 2018. "Bioaccessibility and potential bioavailability of phenolic compounds from achenes as a new target for strawberry breeding programs." *Food Chemistry* 248: 155–65. https://doi.org/10.1016/j.foodchem.2017.11.105

Artursson P., K. Palm, and K. Luthman. 2001. "Caco-2 monolayers in experimental and theoretical predictions of drug transport." *Advanced Drug Delivery Reviews* 46(1–3): 27–43. https://doi.org/10.1016/S0169-409X(00)00128-9

Badyal S., H. Singh, A. K. Yadav, S. Sharma, and I. Bhushan. 2020. "Plant secondary metabolites and their uses." *Plant Archives* 20(2): 3336–40.

Baeza G., B. Sarriá, L. Bravo, and R. Mateos. 2018. "Polyphenol content, *in vitro* bioaccessibility and antioxidant capacity of widely consumed beverages." *Journal of the Science of Food and Agriculture* 98(4): 1397–406. https://doi.org/10.1002/jsfa.8607

Borges G., W. Mullen, A. Mullan, M. E. J. Lean, S. A. Roberts, and A. Crozier. 2010. "Bioavailability of multiple components following acute ingestion of a polyphenol-rich juice drink." *Molecular Nutrition and Food Research* 54(52): S268–77. https://doi.org/10.1002/mnfr.200900611

Caicedo-Lopez L. H., L. Helena, I. Luzardo-Ocampo, M. L. L. Cuellar-Nuñez, R. Campos-Vega, S. Mendoza, and G. Loarca-Piña. 2019. "Effect of the *in vitro* gastrointestinal digestion on free-phenolic compounds and mono/oligosaccharides from *Moringa oleifera* leaves: Bioaccessibility, intestinal permeability and antioxidant capacity." *Food Research International* 120(66): 631–42. https://doi.org/10.1016/j.foodres.2018.11.017

Câmara J. S., B. R. Albuquerque, J. Aguiar, R. C. G. Corrêa, J. L. Gonçalves, D. Granato, J. A. M. Pereira, L. Barros, and I. C. F. R. Ferreira. 2021. "Food ioactive compounds and emerging techniques for their extraction: Polyphenols as a case study." *Foods* 10(1): 37. https://doi.org/10.3390/foods10010037

Campos-Vega R., K. Vázquez-Sánchez, D. López-Barrera, G. Loarca-Piña, S. Mendoza-Díaz, and B. D. Oomah. 2015. "Simulated gastrointestinal digestion and *in vitro* colonic fermentation of spent coffee (*Coffea arabica* L.): Bioaccessibility and intestinal permeability." *Food Research International* 77(2): 156–61. https://doi.org/10.1016/j.foodres.2015.07.024

Carrasco-Sandoval J., M. Aranda-Bustos, K. Henríquez-Aedo, A. López-Rubio, and M. J. Fabra. 2021. "Bioaccessibility of different types of phenolic compounds co-encapsulated in alginate/chitosan-coated zein nanoparticles." *LWT* 149: 112024. https://doi.org/10.1016/j.lwt.2021.112024

Cassani L., A. Gomez-Zavaglia, C. Jimenez-Lopez, C. Lourenço-Lopes, M. A. Prieto, and J. Simal-Gandara. 2020. "Seaweed-based natural ingredients: Stability of phlorotannins during extraction, storage, passage through the gastrointestinal tract and potential incorporation into functional foods." *Food Research International* 137: 109676. https://doi.org/10.1016/j.foodres.2020.109676

Celep E., S. Akyüz, Y. İnan, and E. Yesilada. 2018. "Assessment of potential bioavailability of major phenolic compounds in *Lavandula stoechas* L. ssp. stoechas." *Industrial Crops and Products* 118: 111–7. https://doi.org/10.1016/j.indcrop.2018.03.041

Chen J.-Y., J. Du, M.-L. Li, and C.-M. Li. 2020. "Degradation kinetics and pathways of red raspberry anthocyanins in model and juice systems and their correlation with color and antioxidant changes during storage." *LWT* 128: 109448. https://doi.org/10.1016/j.lwt.2020.109448

Choo K. Y., Y. Y. Ong, R. L. H. Lim, C. P. Tan, and C. W. Ho. 2019. "Study on bioaccessibility of betacyanins from red dragon fruit (*Hylocereus polyrhizus*)." *Food Science and Biotechnology* 28(4): 1163–9. https://doi.org/10.1007/s10068-018-00550-z

Corradini M. G., and M. Peleg. 2007. Shelf-life estimation from accelerated storage data. *Trends in Food Science and Technology* 18(1): 37–47. https://doi.org/10.1016/j.tifs.2006.07.011

D'Antuono I., A. Garbetta, V. Linsalata, F. Minervini, and A. Cardinali. 2015. "Polyphenols from artichoke heads (*Cynara cardunculus* (L.) subsp. scolymus Hayek): *In vitro* bio-accessibility, intestinal uptake and bioavailability." *Food and Function* 6(4): 1268–77. https://doi.org/10.1039/c5fo00137d

Dima C., E. Assadpour, S. Dima, and S. M. Jafari. 2020. "Bioavailability and bioaccessibility of food bioactive compounds; overview and assessment by *in vitro* methods." *Comprehensive Reviews in Food Science and Food Safety* 19(6): 2862–84. https://doi.org/10.1111/1541-4337.12623

Etcheverry P., M. A. Grusak, and L. E. Fleige. 2012. "Application of *in vitro* bioaccessibility and bioavailability methods for calcium, carotenoids, folate, iron, magnesium, polyphenols, zinc, and vitamins B6, B12, D, and E." *Frontiers in Physiology* 3: 317. https://doi.org/10.3389/fphys.2012.00317

Ferreira-Santos P., R. Ibarz, J.-M. Fernandes, A. C. Pinheiro, C. Botelho, C. M. R. Rocha, J. A. Teixeira, and O. Martín-Belloso. 2021. Encapsulated pine bark polyphenolic extract during gastrointestinal digestion: Bioaccessibility, bioactivity and oxidative stress prevention. *Foods* 10(2): 328. https://doi.org/10.3390/foods10020328

Furtado N. A. J. C., L. Pirson, H. Edelberg, L. M. Miranda, C. Loira-Pastoriza, V. Preat, Y. Larondelle, and C. M. André. 2017. "Pentacyclic triterpene bioavailability: An overview of *in vitro* and *in vivo* studies." *Molecules* 22(3): 400. https://doi.org/10.3390/molecules22030400

Herrera-Cazares L. A., A. K. Ramírez-Jiménez, I. Luzardo-Ocampo, M. Antunes-Ricardo, G. Loarca-Piña, A. Wall-Medrano, and M. Gaytán-Martínez. 2021. "Gastrointestinal metabolism of monomeric and polymeric polyphenols from mango (*Mangifera indica* L.) bagasse under simulated conditions." *Food Chemistry* 365: 130528. https://doi.org/10.1016/j.foodchem.2021.130528

Hossain A., and A. Jayadeep. 2020. "Analysis of bioaccessibility of campesterol, stigmasterol, and β-sitosterol in maize by *in vitro* digestion method." *Journal of Cereal Science* 93: 102957. https://doi.org/10.1016/j.jcs.2020.102957

Hoyos-Leyva J. D., A. Chavez-Salazar, F. Castellanos-Galeano, L. A. Bello-Perez, and J. Alvarez-Ramirez. 2018a. "Physical and chemical stability of L-ascorbic acid microencapsulated into taro starch spherical aggregates by spray drying." *Food Hydrocolloids* 83: 143–52. https://doi.org/10.1016/j.foodhyd.2018.05.002

Hoyos-Leyva J. D., L. A. Bello-Pérez, E. Agama-Acevedo, and J. Alvarez-Ramirez. 2018b. "Thermodynamic analysis for assessing the physical stability of core materials microencapsulated in taro starch spherical aggregates." *Carbohydrate Polymers* 197: 431–41. https://doi.org/10.1016/j.carbpol.2018.06.012

Hoyos-Leyva J. D., L. A. Bello-Pérez, and J. Alvarez-Ramirez. 2018. "Thermodynamic criteria analysis for the use of taro starch spherical aggregates as microencapsulant matrix." *Food Chemistry* 259: 175–80. https://doi.org/10.1016/j.foodchem.2018.03.130

Huang Z., Y. Luo, X. Xia, A. Wu, and Z. Wu. 2021. "Bioaccessibility, safety, and antidiabetic effect of phenolic-rich extract from fermented *Psidium guajava* Linn. leaves." *Journal of Functional Foods* 86: 104723. https://doi.org/10.1016/j.jff.2021.104723

Ismail B. B., M. Guo, Y. Pu, O. Çavuş, K. A. Ayub, R. B. Watharkar, T. Ding, J. Chen, and D. Liu. 2021. "Investigating the effect of *in vitro* gastrointestinal digestion on the stability, bioaccessibility, and biological activities of baobab (*Adansonia digitata*) fruit polyphenolics." *LWT* 145: 111348. https://doi.org/10.1016/j.lwt.2021.111348

Ištuk J., D. Šubarić, and L. Jakobek. 2020. "Methodology for the determination of polyphenol bioaccessibility." *Croatian Journal of Food Science and Technology* 12(2): 268–79. https://doi.org/10.17508/CJFST.2020.12.2.16

Jagadeesan G., K. Muniyandi, A. L. Manoharan, G. Nataraj, and P. Thangaraj. 2022. "Understanding the bioaccessibility, α-amylase and α-glucosidase enzyme inhibition kinetics of *Allmania nodiflora* (L.) R. Br. ex Wight polyphenols during *in vitro* simulated digestion." *Food Chemistry* 372: 131294. https://doi.org/10.1016/j.foodchem.2021.131294

Ji S., C. Jia, D. Cao, B. Muhoza, and X. Zhang. 2020. "Formation, characterization and properties of resveratrol-dietary fiber composites: Release behavior, bioaccessibility and long-term storage stability." *LWT* 129: 109556. https://doi.org/10.1016/j.lwt.2020.109556

Karaś M., A. Jakubczyk, U. Szymanowska, U. Złotek, and E. Zielińska. 2017. "Digestion and bioavailability of bioactive phytochemicals." *International Journal of Food Science and Technology* 52(2): 291–305. https://doi.org/10.1111/ijfs.13323

Karuppusamy S. 2009. "A review on trends in production of secondary metabolites from higher plants by *in vitro* tissue, organ and cell cultures." *Journal of Medicinal Plants Research* 3(13): 1222–39.

Krishnan G., J. Subramaniyan, P. C. Subramani, B. Muralidharan, and D. Thiruvengadam. 2017. "Hesperetin conjugated PEGylated gold nanoparticles exploring the potential role in anti-inflammation and anti-proliferation during diethylnitrosamine-induced hepatocarcinogenesis in rats." *Asian Journal of Pharmaceutical Sciences* 12(5): 442–55. https://doi.org/10.1016/j.ajps.2017.04.001

Kuck L. S., J. L. Wesolowski, and C. P. Z. Noreña. 2017. "Effect of temperature and relative humidity on stability following simulated gastro-intestinal digestion of microcapsules of Bordo grape skin phenolic extract produced with different carrier agents." *Food Chemistry* 230: 257–64. https://doi.org/10.1016/j.foodchem.2017.03.038

Lassoued M. A., F. Khemiss, and S. Sfar. 2011. "Comparative study of two *in vitro* methods for assessing drug absorption: Sartorius SM 16750 apparatus versus everted gut sac." *Journal of Pharmacy and Pharmaceutical Sciences* 14(1): 117–27. https://doi.org/10.18433/J3GC7R

Liu J., J. Bi, X. Liu, D. Liu, X. Wu, J. Lyu, and Y. Ding. 2020. "Effects of pectins and sugars on β-carotene bioaccessibility in an *in vitro* simulated digestion model." *Journal of Food Composition and Analysis* 91: 103537. https://doi.org/10.1016/j.jfca.2020.103537

Liu W., J. Wang, D. J. McClements, and L. Zou. 2018. "Encapsulation of β-carotene-loaded oil droplets in caseinate/alginate microparticles: Enhancement of carotenoid stability and bioaccessibility." *Journal of Functional Foods* 40: 527–35. https://doi.org/10.1016/j.jff.2017.11.046

Luzardo-Ocampo I., G. Loarca-Piña, and E. Gonzalez de Mejia. 2020. "Gallic and butyric acids modulated NLRP3 inflammasome markers in a co-culture model of intestinal inflammation." *Food and Chemical Toxicology* 146: 111835. https://doi.org/10.1016/j.fct.2020.111835

Luzardo-Ocampo I., R. Campos-Vega, E. G. de Mejia, and G. Loarca-Piña. 2020a. "Consumption of a baked corn and bean snack reduced chronic colitis inflammation in CD-1 mice via downregulation of IL-1 receptor, TLR, and TNF-α associated pathways." *Food Research International* 132: 109097. https://doi.org/10.1016/j.foodres.2020.109097

Luzardo-Ocampo I., A. K. Ramírez-Jiménez, Á. H. Cabrera-Ramírez, N. Rodríguez-Castillo, R. Campos-Vega, G. Loarca-Piña, and M. Gaytán-Martínez. 2020b. "Impact of cooking and nixtamalization on the bioaccessibility and antioxidant capacity of phenolic compounds from two sorghum varieties." *Food Chemistry* 309: 125684. https://doi.org/10.1016/j.foodchem.2019.125684

Luzardo-Ocampo I., R. Campos-Vega, M. Gaytán-Martínez, R. Preciado-Ortiz, S. Mendoza, G. Loarca-Piña. 2017. "Bioaccessibility and antioxidant activity of free phenolic compounds and oligosaccharides from corn (*Zea mays* L.) and common bean (*Phaseolus vulgaris* L.) chips during *in vitro* gastrointestinal digestion and simulated colonic fermentation." *Food Research International* 100(1): 304–11. https://doi.org/10.1016/j.foodres.2017.07.018

Mateer S. W., J. Cardona, E. Marks, B. J. Goggin, S. Hua, and S. Keely. 2016. "*Ex vivo* intestinal sacs to assess mucosal permeability in models of gastrointestinal disease." *Journal of Visualized Experiments* 108: e53250. https://doi.org/10.3791/53250

McClements D. J. 2020. "Advances in nanoparticle and microparticle delivery systems for increasing the dispersibility, stability, and bioactivity of phytochemicals." *Biotechnology Advances* 38: 107287. https://doi.org/10.1016/j.biotechadv.2018.08.004

McClements D. J., and S.-F. Peng. 2020. "Current status in our understanding of physicochemical basis of bioaccessibility." *Current Opinion in Food Science* 31: 57–62. https://doi.org/10.1016/j.cofs.2019.11.005

Montiel-Sánchez M., T. García-Cayuela, A. Gómez-Maqueo, H. S. García, and M. P. Cano. 2021. "*In vitro* gastrointestinal stability, bioaccessibility and potential biological activities of betalains and phenolic compounds in cactus berry fruits (*Myrtillocactus geometrizans*)." *Food Chemistry* 342: 128087. https://doi.org/10.1016/j.foodchem.2020.128087

Neilson A. P., T. N. Sapper, E. M. Janle, R. Rudolph, N. V. Matusheski, and M. G. Ferruzzi. 2010. "Chocolate matrix factors modulate pharmacokinetic behavior of cocoa flavan-3-ol phase-II metabolites following oral consumption by Sprague-Dawley rats." *Journal of Agricultural and Food Chemistry* 58(11): 6685–91. https://doi.org/10.1021/jf1005353

Nunes R., C. Silva, and L. Chaves. 2016. "Tissue-based *in vitro* and *ex vivo* models for intestinal permeability studies." In *Concepts and Models for Drug Permeability Studies*, edited by B. Sarmento, 203–36. Elsevier. https://doi.org/10.1016/B978-0-08-100094-6.00013-4

Ovando-Martínez M., N. Gámez-Meza, C. C. Molina-Domínguez, C. Hayano-Kanashiro, and L. A. Medina-Juárez. 2018. "Simulated gastrointestinal digestion, bioaccessibility and antioxidant capacity of polyphenols from red chiltepin (*Capsicum annuum* L. var. *glabriusculum*) grown in northwest Mexico." *Plant Foods for Human Nutrition* 73(2): 116–21. https://doi.org/10.1007/s11130-018-0669-y

Palafox-Carlos H., J. F. Ayala-Zavala, and G. A. González-Aguilar. 2011. "The role of dietary fiber in the bioaccessibility and bioavailability of fruit and vegetable antioxidants." *Journal of Food Science* 76(1): R6–15. https://doi.org/10.1111/j.1750-3841.2010.01957.x

Parada J., and J. M. Aguilera. 2007. "Food microstructure affects the bioavailability of several nutrients." *Journal of Food Science* 72(2): R21–R32. https://doi.org/10.1111/j.1750-3841.2007.00274.x

Perestrelo R., E. Rodriguez, and J. S. Câmara. 2017. "Impact of storage time and temperature on furanic derivatives formation in wines using microextraction by packed sorbent tandem with ultrahigh pressure liquid chromatography." *LWT-Food Science and Technology* 76: 40–7. https://doi.org/10.1016/j.lwt.2016.10.041

Putnik P., D. Gabrić, S. Roohinejad, F. J. Barba, D. Granato, K. Mallikarjunan, J. M. Lorenzo, and D. B. Kovačević. 2019. "An overview of organosulfur compounds from *Allium* spp.: From processing and preservation to evaluation of their bioavailability, antimicrobial, and anti-inflammatory properties." *Food Chemistry* 276: 680–91. https://doi.org/10.1016/j.foodchem.2018.10.068

Qi W., Z. Zhang, and T. Wu. 2020. "Encapsulation of β-carotene in oleogel-in-water pickering emulsion with improved stability and bioaccessibility." *International Journal of Biological Macromolecules* 164: 1432–42. https://doi.org/10.1016/j.ijbiomac.2020.07.227

Rein M. J., M. Renouf, C. Cruz-Hernandez, L. Actis-Goretta, S. K. Thakkar, and M. da S. Pinto. 2013. "Bioavailability of bioactive food compounds: A challenging journey to bioefficacy." *British Journal of Clinical Pharmacology* 75(3): 588–602. https://doi.org/10.1111/j.1365-2125.2012.04425.x

Ribas-Agustí A., O. Martín-Belloso, R. Soliva-Fortuny, and P. Elez-Martínez. 2018. "Food processing strategies to enhance phenolic compounds bioaccessibility and bioavailability in plant-based foods." *Critical Reviews in Food Science and Nutrition* 58(15): 2531–48. https://doi.org/10.1080/10408398.2017.1331200

Ribeiro V. R., G. M. Maciel, M. M. Fachi, R. Pontarolo, I. de A. A. Fernandes, A. P. Stafussa, and C. W. I. Haminiuk. 2019. "Improvement of phenolic compound bioaccessibility from yerba mate (*Ilex paraguariensis*) extracts after biosorption on *Saccharomyces cerevisiae*." *Food Research International* 126: 108623. https://doi.org/10.1016/j.foodres.2019.108623

Rousseau S., C. Kyomugasho, M. Celus, M. E. G. Hendrickx, and T. Grauwet. 2020. "Barriers impairing mineral bioaccessibility and bioavailability in plant-based foods and the perspectives for food processing." *Critical Reviews in Food Science and Nutrition* 60(5): 826–43. https://doi.org/10.1080/10408398.2018.1552243

Sánchez-Rangel J. C., J. Benavides, J. B. Heredia, L. Cisneros-Zevallos, and D. A. Jacobo-Velázquez. 2013. "The Folin–Ciocalteu assay revisited: Improvement of its specificity for total phenolic content determination." *Analytical Methods* 5(21): 5990. https://doi.org/10.1039/c3ay41125g

Scholz S., and G. Williamson. 2007. "Interactions affecting the bioavailability of dietary polyphenols *in vivo*." *International Journal for Vitamin and Nutrition Research* 77(3): 224–35. https://doi.org/10.1024/0300-9831.77.3.224

Schramm D. D., M. Karim, H. R. Schrader, R. R. Holt, N. J. Kirkpatrick, J. A. Polagruto, J. L. Ensunsa, H. H. Schmitz, and Carl L. Keen. 2003. "Food effects on the absorption and pharmacokinetics of cocoa flavanols." *Life Sciences* 73(7): 857–69. https://doi.org/10.1016/S0024-3205(03)00373-4

Shahidi F., and H. Peng. 2018. "Bioaccessibility and bioavailability of phenolic compounds." *Journal of Food Bioactives* 4: 11–68. https://doi.org/10.31665/JFB.2018.4162

Shahidi F., and Y. Pan. 2021. "Influence of food matrix and food processing on the chemical interaction and bioaccessibility of dietary phytochemicals: A review." *Critical Reviews in Food Science and Nutrition* 62(23): 6421–45. https://doi.org/10.1080/10408398.2021.1901650

Świeca M., U. Gawlik-Dziki, D. Dziki, and B. Baraniak. 2017. "Wheat bread enriched with green coffee - *In vitro* bioaccessibility and bioavailability of phenolics and antioxidant activity." *Food Chemistry* 221: 1451–57. https://doi.org/10.1016/j.foodchem.2016.11.006

Szabo K., B. E. Teleky, F. Ranga, E. Simon, O. L. Pop, V. Babalau-Fuss, N. Kapsalis, and D. C. Vodnar. 2021. "Bioaccessibility of microencapsulated carotenoids, recovered from tomato processing industrial by-products, using *in vitro* digestion model." *LWT* 152: 112285. https://doi.org/10.1016/j.lwt.2021.112285

Tian W., R. Hu, G. Chen, Y. Zhang, W. Wang, and Y. Li. 2021. "Potential bioaccessibility of phenolic acids in whole wheat products during *in vitro* gastrointestinal digestion and probiotic fermentation." *Food Chemistry* 362: 130135. https://doi.org/10.1016/j.foodchem.2021.130135

Tomas M., L. Zhang, G. Zengin, G. Rocchetti, E. Capanoglu, and L. Lucini. 2021. "Metabolomic insight into the profile, *in vitro* bioaccessibility and bioactive properties of polyphenols and glucosinolates from four Brassicaceae microgreens." *Food Research International* 140: 110039. https://doi.org/10.1016/j.foodres.2020.110039

Tomsone L., R. Galoburda, Z. Kruma, and K. Majore. 2020. "Physicochemical properties of biscuits enriched with horseradish (*Armoracia rusticana* L.) products and bioaccessibility of phenolics after simulated human digestion." *Polish Journal of Food and Nutrition Sciences* 70(4): 419–28. https://doi.org/10.31883/pjfns/130256

Vega-Rojas L. J., I. Luzardo-Ocampo, J. Mosqueda, D. M. Palmerín-Carreño, A. Escobedo-Reyes, A. Blanco-Labra, K. Escobar-García, and T. García-Gasca. 2021. "Bioaccessibility and *in vitro* intestinal permeability of a recombinant lectin from tepary bean (*Phaseolus acutifolius*) using the everted intestine assay." *International Journal of Molecular Sciences* 22(3): 1049. https://doi.org/10.3390/ijms22031049

Yang M., S. I. Koo, W. O. Song, and O. K. Chun. 2011. "Food matrix affecting anthocyanin bioavailability: Review." *Current Medicinal Chemistry* 18 (2): 291–300. https://doi.org/10.2174/092986711794088380

Zhang B., Y. Zhang, H. Li, Z. Deng, and R. Tsao. 2020. "A review on insoluble-bound phenolics in plant-based food matrix and their contribution to human health with future perspectives." *Trends in Food Science and Technology* 105: 347–62. https://doi.org/10.1016/j.tifs.2020.09.029

Zhang H., G. Liu, X. Zeng, Y. Wu, C. Yang, L. Mei, Z. Wang, and L. Huang. 2015. "Fabrication of genistein-loaded biodegradable TPGS-b-PCL nanoparticles for improved therapeutic effects in cervical cancer cells." *International Journal of Nanomedicine* 10: 2461–73. https://doi.org/10.2147/IJN.S78988

Zorenc Z., R. Veberic, F. Stampar, D. Koron, and M. Mikulic-Petkovsek. 2017. "Thermal stability of primary and secondary metabolites in highbush blueberry (*Vaccinium corymbosum* L.) purees." *LWT-Food Science and Technology* 76: 79–86. https://doi.org/10.1016/j.lwt.2016.10.048

Index

A

B

C

D

H

I

J

K

L

Q

R

S

T

U

V